PROCESSES AND BOUNDARIES OF THE MIND

Contemporary Systems Thinking

Series Editor: Robert L. Flood
Monash University
Australia

CRITICAL SYSTEMIC PRAXIS FOR SOCIAL AND ENVIRONMENTAL JUSTICE
Participatory Policy Design and Governance for a Global Age
Janet McIntyre-Mills

DESIGNING SOCIAL SYSTEMS IN A CHANGING WORLD
Bela H. Banathy

GUIDED EVOLUTION OF SOCIETY
A Systems View
Bela H. Banathy

METADECISIONS
Rehabilitating Epistemology
John P. van Gigch

POWER, IDEOLOGY, AND CONTROL
John C. Oliga

PROCESSES AND BOUNDARIES OF THE MIND
Extending the Limit Line
Yair Neuman

SOCIOPOLITICAL ECOLOGY
Human Systems and Ecological Fields
Frederick L. Bates

SYSTEMIC INTERVENTION
Philosophy, Methodology, and Practice
Gerald Midgley

SYSTEMS METHODOLOGY FOR THE MANAGEMENT SCIENCES
Michael C. Jackson

A Continuation Order Plan is available for this series. A continuation order will bring delivery of each new volume immediately upon publication. Volumes are billed only upon actual shipment. For further information please contact the publisher.

PROCESSES AND BOUNDARIES OF THE MIND

Extending the Limit Line

Yair Neuman
Ben-Gurion University of the Negev
Beer-Sheva, Israel

Kluwer Academic / Plenum Publishers
New York, Boston, Dordrecht, London, Moscow

ISBN: 0-306-48121-9

©2003 Kluwer Academic / Plenum Publishers
233 Spring Street, New York, New York 10013

http://www.wkap.nl

10 9 8 7 6 5 4 3 2 1

A C.I.P. record for this book is available from the Library of Congress

All rights reserved

No part of this book may be reproduced, stored in a retrieval system, or transmitted in any form or by any means, electronic, mechanical, photocopying, microfilming, recording, or otherwise, without written permission from the Publisher, with the exception of any material supplied specifically for the purpose of being entered and executed on a computer system, for exclusive use by the purchaser of the work.

Permissions for books published in Europe: *permissions@wkap.nl*
Permissions for books published in the United States of America: *permissions@wkap.com*

Printed in the United Kingdom by Biddles/IBT Global

To Orna, Yiftach, Yaara and Tamar

and

To Zvi Bekerman, who first introduced me to the writings of Bateson, Bakhtin and Volosinov

Preface

Men who love wisdom must be good inquirers into many things indeed.
                                                                Heraclitus

Someone has amusingly defined an intellectual as a person who finds at least one thing more interesting than sex. As a young child, before sex was an appealing alternative, I found myself fascinated by ideas. While my schoolmates were obsessively occupied with soccer and collecting bubblegum cards, I was occupied with collecting bubblegum cards *and* reading books that caught me in their imaginary world. I can still remember myself as a schoolboy playing with ideas and for instance trying passionately to convince a friend that, based on Darwin's theory of evolution, chimpanzees all over the world were going to evolve into an intelligent species and would get even with humankind for treating them so badly. Since my naive theory of evolution was based on pseudo-scientific comic books, I assumed that men had evolved from chimps, and I considered it only a matter of time until chimps would evolve into human beings. Today, I am amused to find that this naive theory of Darwinian evolution is still evident among fundamentalist preachers and laymen alike. Throughout the following years, my knowledge of Darwin's theory has much improved, although my moral stance concerning chimps and humankind has remained the same. But at that time, it is unnecessary to add, my friend considered my idea totally foolish and refuted my scholarly speculations by assuring me that the only chimp that he was familiar with lived in the local zoo under the tight scrutiny of armed guardians. This was quite a shock for my naive theory of evolution, but it did not prevent me from continuing to play with ideas and being fascinated by them, whatever their validity.

Lucky enough to survive the anti-intellectual atmosphere of formal education, I can testify that I am still fascinated by ideas. My wish is that this book will be read as an experience in evoking the same childish fascination of ideas, as an experimentation[1] in ideas, and not as a tiresome, scholarly treatise.

Acknowledgments

This manuscript is the expression of a long and continuing process of intellectual inquiry and development. I had the privilege of conducting this voyage in the company of several dear friends who supported me all the way. I am grateful to Zvi Bekerman, Irun Cohen, Joe Engelberg, Peter Harries-Jones, and Jerry Unger for their intellectual and constructive support of my work

The publication of this book was supported by two grants from the Faculty of Humanities and Social Sciences, Ben-Gurion University of the Negev. I would especially like to thank Prof. Jimmy Weinblat for his kind support in the publication of this book.

I would like to thank the following publishers for their permission to reprint copyrighted materials: Taylor and Francis Books for reprinting excerpts from M. Merleu-Ponty (1962), *Phenomenology of Perception*; Harvard University Press for reprinting excerpts from A. R. Luria (1976), *Cognitive Development: Its Cultural and Social Foundation*; John Wiley and Sons, Inc., for reprinting Y. Neuman (2001), On Turing's Carnal Error: Some Guidelines for a Contextual Inquiry into the Embodied Mind, *Systems Research and Behavioral Sciences*, **18**, 557–565; University of Texas Press for reprinting excerpts from M. M. Bakhtin (1990), *Art and Answerability: Early Philosophical Essays*, edited by Michael Holquist and Vadim Liapunov, translation and notes by Vadim Liapunov; Academic Press for reprinting excerpts from V. N. Volosinov (1998), *Marxism and the Philosophy of Language*.

I would like to close by thanking Robert Flood for inviting me to publish this manuscript in the Contemporary Systems Thinking series, and Henry Gomm and Ken Derham for their editorial support.

Contents

1.	Introduction	1
2.	On What There Is	5
3.	In the Beginning was the Act	9
	Cat-logue 1: The Cheshire Cat and Descartes	15
4.	Beyond the Curtain or into the Looking Glass	17
5.	A Guided Tour in the Kingdom of Signs	25
	Cat-logue 2: Are There Jewish Cats?	37
6.	Saussure and Semiotics as a Social System	41
7.	The Mind as a Semiotic Interface	47
	Cat-logue 3: It Means Nothing	53
8.	We Have Never Been Too Abstract	55
	Cat-logue 4: Where Does the Frame End?	65
9.	A Snake that Bites Its Tail	67
10.	The Demon of Circularity	79
11.	Origins	83
	Cat-logue 5: The Hole in the Bagel	87
12.	Laws of Form	89
	Cat-logue 6: Inside the Outside	97
13.	Toward a Phenomenology of Boundaries	99
	Cat-logue 7: How Deep is the Surface?	105
14.	Peter Pan's Shadow and the Empty Observer	107
15.	On Turing's Carnal Error	115

16.	What is so Complex about Complexity?	125
	Cat-logue 8: Speaking with the Cat about Spinoza	131
17.	Toward a Dialogical Complexity	139
18.	The Architectonics of the Mind	143
	Cat-logue Which Is an Epilogue: Where a Blind Man Ends	157

References	159
Endnotes	163
Name Index	167
Subject Index	169

Chapter 1
Introduction

The wish to write a book that evokes fascination with ideas largely shaped the character of the book, from the arrangement of its general themes to its particular style. Although it is a scholarly treatise, written according to the accepted norms of academic rhetoric, it includes several aspects that differentiate it from the common academic monograph. For example, the book includes imaginary dialogues with my cat ("cat-logues") that aim to explore further the ideas presented in the chapters, from a humorous and reflective position. These cat-logues are a homage to Gregory Bateson's famous metalogues with his daughter. In contrast to Bateson, my children are used to their father's ideas and therefore the dramatic impact evident in Bateson's dialogue with his naive daughter cannot be repeated. Therefore, I chose my white cat "Bamba" (named for a popular Israeli snack), who is the true intellectual in our house, to be my partner for those conversations. However, there is another reason for choosing a cat as an intellectual partner for scholarly dialogue. As will later be presented in this book, several philosophers, such as Spinoza and Bergson, considered intuition the highest form of knowledge. Intuition as the ability to grasp the specific modes of reality in the context of the ultimate whole is evident in the life of animals.[2] Animals in general and cats in particular live in what Merleau-Ponty describes as a *pre-objective* form of being. They are *being-in-the-world* as an integral part of it, without the ability, or the burden, to reflect on their existence through questions such as "Why is there something instead of nothing?" This primordial form of being clearly involves intuition rather than scholarly reflection. Since this book involves an inquiry into this pre-objective reality as the origin of our mind, the decision to choose a cat as my "alter ego" for discussing philosophical questions and the nature of being from a fresh perspective seemed a natural choice. These cat-logues do not dismiss the conventional form of scientific rhetoric. I support the ideas presented in the book with an arsenal of arguments that are supposed to defend my thesis. However, although I consider argumentation to be the *sine qua non* (a Latin expression which is itself an example of academic rhetoric!) of the humanities and the social sciences, I do not believe that we should confuse this process of argumentation with the intellectual work in itself.[3] It is a common practice, particularly among professional academics, to repeatedly make this error by focusing on the quality of the arguments provided in favor of a thesis, rather than on the quality of the thesis or the quality of the experimentation process in which they are involved. Scientific work, or what may be better called "intellectual work," involves primarily curiosity and ideas. As Deleuze has already argued, the main task of philosophy is the generation of new ideas through which we can examine events in a fresh way. This is exactly the aim of this book.

A process of argumentation that justifies and contests may provide ideas. However, good ideas have some kind of aesthetic value that places them in a sacred position beyond the scope of academic rhetoric. I am not naive enough to consider the differentiation between ideas and rhetoric as possible or easy, but I am naive enough to support the primacy of ideas over academic rhetoric. This is not to say that I do not support my thesis with arguments. On the contrary! As someone who is also involved in the study of discourse, rhetoric and argumentation, I recruited my entire rhetorical arsenal (and professional jargon) in order to present my ideas in the most impressive way. However, I would like the reader to read beyond my rhetoric, and beyond the references to previous scholarly work and the "big shots" of the past and to judge how *interesting* are the ideas presented in this book.

The idea that scientific ideas should be interesting rather than "true," "empirically valid" or even genuine (a phantasy of the modern mind which denies its embodiment in history) may sound like fashionable nonsense provided by some kind of young post-modernist, or post-post modernist writer. This image is wrong and there is no better (rhetorically better) way of dismissing it than by quoting a prominent scientist, who holds the same position.

In the preface to his book, "Mathematical models of morphogenesis," Rene Thom (1983) writes:

The reader will consequently be in a position to ask himself whether the considerations offered here are susceptible to experimental verification, or even whether they are true. I must admit that the problem of truth has not concerned me directly. I am, however, convinced of one point: as well as the truth of a theory, or a model, one must also consider its *interest*. If we are to believe Karl Popper, psychoanalysis is not "falsifiable," and hence must be placed outside science. And yet psychoanalysis offers infinitely more interest than many scientific theories whose truth is indisputably assured. It is in this spirit that I offer here these models, not so much as testable hypotheses or as experimentally controllable models but as a *stimulus to the imagination which leads to the exercise of thought and this is an increase in our understanding of the world and man.* [my emphasis]

Thom's statement seems "non-scientific" to those who still believe that the spirit of science is data gathering and hypotheses testing. How does he dare to state publicly that the question of truth does not bother him directly? To this audience, Thom's statement and the ideas presented in this book may cause some slight physical anxiety.

Nevertheless, every physical and intellectual growth involves some pain as schemes and structures are forced to change. Excluding pathological cases, there is no biological organ forced by the body not to grow. Surprisingly, human systems have myriad counter examples, and the social sciences sometimes present a strong academic resistance to this growth. My own personal experience provides me with several examples. Once, I presented a lecture to a group of researchers from education and the behavioral sciences entitled: "Star worship." This lecture concerned the most popular method of statistical hypotheses testing in the behavioral sciences: statistical significance testing and its abuse by "social scientists." Anyone who has opened a journal on psychology, education or sociology is familiar with the stars covering the research papers and allegedly providing the hypotheses of the researcher: "This result was found statistically significant ($p = .05$ marked by '*')." In contrast to the scientific sex appeal of those tests, there is a sufficient body of knowledge showing that this senseless practice that governs the behavioral sciences, and provides them with its "scientific" cloth, adds nothing to the development of the behavioral sciences and even hinders its development. The unavoidable conclusion is that this practice in its current form should be abandoned in order to free the study of the human and to allow it to be developed in a better way. This is the ideal of a scientific inquiry. After we learn that a certain practice is prohibiting the advancement of knowledge (a la Francis Bacon), we should abandon this practice and look for better ways to conduct our inquiry. In practice, we are all familiar with the sociological, psychological and other ideological systems that mediate the quest for knowledge and hinder its development. In my specific case, most of the reactions I received to my lecture ranged from "Ad Hominem" arguments, that is, arguments against the person rather than against the argument ("Do you think that you are smarter than the researchers in the field?"), of those who understand nothing about the mathematical-statistical model but trust the gurus of the field (that is, journal editors), to those who claim to understand the mathematical statistical model of statistical significance testing, but request to accept the norms of research since those are "the rules of the game." For those who do not want to accept the rules of the game, open-minded invitations for intellectual work, such as those presented above by Thom, may be a major source of encouragement.

THE AUDIENCE

This book is about the way we know the world and ourselves and about an alternative way through which we may know the world and ourselves. In this sense, the book should be of great interest to various audiences such as: cognitive scientists, semioticians, psychologists, philosophers, educators, linguists, and members of the systems science community who usually feel at home everywhere as long as the intellectual discussion does not slip into professional jargon and technical detail. In order to open the door for different audiences, to include the general, educated reader, I have made a substantial effort to avoid professional jargon and technical details, and open each chapter with a brief abstract that summarizes its main points. I am well aware that at certain loci of my book the discussion drops into professional rhetoric, which I found extremely difficult to avoid. The reader is expected to be patient with this language and to understand that what may be experienced as obscure language is sometimes a struggle to constitute a new way of thinking about old problems saturated in their own jargon and rhetoric.

FINAL COMMENTS

The book propagates a process-oriented perspective of mind-reality and, therefore, questions the ontological status of formal, static forms we impose on reality. The implications of this intellectual venture are evident not only in the conclusions I draw, but also in the style that may shift whimsically from an analytic form of argumentation to poetic language. Therefore, the reader who expects coherence in terms of style and content might be disappointed. She should bear in mind that coherence is a property of artificial systems and not of the human mind. In this sense, systematization is beyond the scope of this book, which is only experimentation with ideas.

Another point is that as a proponent of the systems science venture, with its wide intellectual scope, I could not resist discussing a variety of issues in various fields: from semiotics to psychology, from Spencer-Brown's calculus of indication to Rene Magritte's famous painting, "This is not a pipe," from cultural psychology to the concept of nothingness in mysticism. Therefore, the book draws on several major domains, such as semiotics, systems research, epistemology and cognition. This intellectual result may challenge some readers, but might worry others, especially those who are still captivated by the artificial academic demarcation of domains and professions. Those readers may find themselves again and again trying to figure out what academic niche this book occupies. This problem may especially bother librarians. I can imagine them anxiously searching the Dewey manual in order to decide where to place this book. "It is a book in the System Thinking series, therefore it belongs in the science section, just mark it Q 141.N12." "Hey, but this book mainly deals with semiotics! So let us put it in the linguistic or the philosophy section." "No! It is a book about epistemology, so..." and so on. This kind of anxiety should not be underestimated, since it reflects our existential anxiety from encountering Being by abandoning our mediated and fragmented thinking. The answer to this perplexity, and to the librarians' agony, that is the librarian's Heidegerian *Angst*, is expressed in the body of the text, as the reader will hopefully find out. This book is about the dynamic nature of the human mind and this dynamic is not the sole property of a specific academic discipline.

Chapter 2
On What There Is

Some years ago now I observed the multitude of errors that I had accepted as true in my earliest years, and the dubiousness of the whole superstructure I had since reared on them.
Descartes

> Summary: The world as it appears to us is a world of objects, whether concrete or abstract. This is the reified universe. It is a question whether our world is "really" populated by objects, or whether those objects are static forms imposed on reality through our mind.

Any serious quest for knowledge should start with some dissatisfaction with an established representation of the world, which usually rests peacefully unnoticed, feigning to be reality itself. I do not mean that any serious quest for knowledge must be agnostic, but that some kind of general dissatisfaction should accompany the tool kit of science or any other form of intellectual inquiry. This skeptical stance is a tactical one and not an aim for itself. Sometimes critical thinkers, specifically those who are associated with the "post-modern" venture, express great pleasure at showing that our representations of the world rest on quicksand rather than on solid ground. However, without having a constructive strategy that provides an alternative to the idols of our mind, such a critical inquiry may end in both scientific and moral decadence. Indeed, Emmanuel Kant already criticized the skeptics by comparing them to nomadic tribes that attack the established cities without maintaining cities of their own.[4]

The aim of this book is to inquire into one of the most salient representations of the Western mind. For didactic reasons I will not present this representation now, but invite you to begin our quest from a different place.

The place I want to start is a personal confession. As someone who has been educated by experimental psychologists, I was always told that we could not uncover general rules of the mind by using case studies and introspection. We cannot uncover rules of the mind by (1) trusting the results of a single case over the results of a representative sample, and (2) by using the subject's report of his "internal" states rather than by counting on observable data. In fact, when I taught research methods to university students, I used to open the first lesson by criticizing (and mocking) those primitive sources of knowledge and by contrasting them with the merits of the "scientific" methodology in the behavioral sciences.[5]

I would like to invite you, as a reader of this book, to participate in a brief *Gedanken Study* (thought experiment) that aims to refute my past positivistic arrogance. The instructions for this experiment are apparently very simple: Think for a minute and then provide me with a description of what exists in the world or "what there is." This task is not limited to your observable world, but may also involve abstract entities that populate your "internal" world, which is not directly available to outsiders. At this phase of our inquiry, please do not ask for further information or clarification for the experiment. Consider it as a phenomenological experiment in which you are studying the essential structures of pure consciousness as unmediated experience.[6] Unfortunately, I cannot be evident during this experimentation. However, I hypothesize that your introspective case study would result in a list of *things*: People, cats, a chair, numbers, etc. If my hypothesis is confirmed, as I suspect, then through an introspective case study you just revealed a general rule of the mind:

Our world is conceived as almost exclusively populated by things-objects, whether concrete objects or abstracts.

At this phase of our inquiry, I do not want to get into an elaborate discussion of what it means to be an object/thing,[7] and my suggestion is to consider this concept as pre-theoretically meaningful. If we accept this position, then I believe that no one can argue against the scientific validity of your case study's conclusion. You see how simple it is to uncover universal rules! Just look inside. However, this "rule" may sound to those people who hold a realistic stance (that is people who think that the world exists separate from our mind and preceding our consciousness) as trivial in a best-case scenario, and totally foolish in a worse-case scenario. After all, what is there in our world except for things?!

This question necessarily shifts our discussion from the world to the mind, since people (at least at the current phase of the Western culture) reflect on their world as almost exclusively inhabited by objects, and this phenomenon is even interpreted as a basic and "universal rule" of the mind. Indeed, Piaget argues that our ability to perceive permanent objects (the object continuity phenomenon[8]) is one of the basic and primary phases of our cognitive development. However, things are much more intricate since the evidence that people in the current phase of the Western culture *describe* (or reflect upon) their world as almost exclusively populated by objects is by no means valid scientific evidence regarding the structure of reality. In other words, is this reified conception of reality reality in itself? Is the map the territory? Or is it only one possible and specific representation of reality (psychological or metaphysical) that pretends to be reality in itself?

I opened this section by suggesting that any serious quest for knowledge should start with dissatisfaction with a well-established representation of reality. The idea that our world is almost exclusively populated by things/objects (the "Reified Universe") is one of those well-established representations I would like to seek. Recalling Kant's criticism, I would also like to offer a constructive reading of the reified universe phenomenon, and to present the idea that *objects* do not precede mind but are the phenomenological expression of *mind-reality* synergism. I deliberately use the term "mind-reality" as expressing a theoretical stance suggesting that mind and reality are coupled, and should be regarded as a mysterious encounter that results in a world of appearances. By using this term, I clearly adhere to the phenomenological theory of Merleau-Ponty that aims to transcend classical dualities, such as the mind-body duality or the mind-world duality. I further suggest that when we adopt this perspective, many old and difficult problems concerning the nature of mind-reality, body-mind, language-world, and the way we know the world, may be seen in a different light. More specifically, this book suggests that the objects that we think populate our world are in fact processes of signification that have been the subject of a reified metaphysics. My aim is to describe this reification by inquiring into the mind as a socio[9]-somatic-semiotic motif and the way it emerges from its processual[10] and primordial origins. In this sense, one possible (and surprisingly positive) outcome of this book is the *"reenchantment of the mind,"*[11] by liberating it from its captivity in a reified metaphysics.

Since my past intellectual arrogance has already been exposed, I feel obliged at the end of this chapter to present a more modest intellectual position by acknowledging my debt to previous scholars. Many (if not all) of the ideas presented in this book have already appeared, although in different forms, in the writings of Bateson, Volosinov, Bakhtin, Merleau-Ponty, Peirce and other radical voices. Although I do not adhere to any specific theory, I use those ideas as the building blocks of this book. The only thing that is new is their combination. However, is not that what language is all about? A unique and creative combination that emerges out of a finite set of components (ideas in my case) and syntax? In this sense, the reader should not expect this book to reveal to him anything new since language when it turns to inquire into its own origins never tells us something new, but the most trivial we have forgotten.

Chapter 3
In the Beginning was the Act

The work of the mind exists only in act.
Valery

> Summary: Although we see our world as populated by objects, a close examination of our most basic form of being-in-the-world suggests that objects simply do not exist. At the most basic form of existence, we encounter "dynamic objects," singularities in their flux of being.

In the previous chapter I presented the idea that although the world appears to us to be populated by objects, there is another way to interpret this phenomenon, rather than to assume that reality is reified. This suggestion does not nullify or empty the world of objects. We cannot deny that objects exist in our phenomenological world. We eat those things, we see those things, and we touch those things and we can never say a thing or think about some *thing* without their presence. So what exactly is the point in questioning their appearance? Well, the idea is not to question the appearances, but to question the meaning of those appearances and whether they actually correspond to real objects. The idea is that on the most basic level of encountering the world there are no objects. We simply cannot find them. Following this still unexplained suggestion, an "object" may be considered as a certain dynamic (what Peirce describes as the *Dynamic Object*), which during its coupling with an organism has somehow been transformed into a stable structure. In order to explain this suggestion, let us closely examine the most basic existential state, our most basic form of being-in-the-world.

In our experience with the world we operate and interact with what we *reflect* upon as concrete and particular objects: This computer, that cat, etc. I use the expression "reflect upon" because when we first interact with "objects" in our environment we do not encounter them in a reflective, analytic way, but in a non-reflective manner. That is, our bodies correspond to certain mysterious events in reality, but these events are not yet objects. Objects do not seem to exist as consummated and fixed entities that precede mind. This suggestion raises the question what precedes what: mind or reality? After all, the world of appearances is the result of mental activity that differentiates a given universe (reality) into categories: cats, mice, apples, etc. Therefore, it seems that mind precedes the world. Nevertheless, mind exists only by approaching certain portions of the world. There is no mind without assuming a world which this mind heeds. Aha! Therefore, the world precedes mind! Both ontological positions are deficient since they get us into a regression of mind-world we should definitely avoid. A possible solution to this perplexity is by transcending the old discourse that sharply separates between mind and world.

We may adopt the position that the mind and the world come into being at their point of intersection. According to this suggestion our basic unit of analysis should be neither the world nor the mind, but a certain form of interaction (what Vygotsky called *Activity*) that constitutes the mind and the world as two different phenomena. The idea that mind and world come into being as interaction (what will be later described as semiosis) brings us to the conception of *Lebensphilosophie* (philosophy of life) advocated by Nietzsche, Dilthey and others. Lebensphilosophie conceives of life as a process of *becoming* and points at our inability to grasp this process in its ultimate purity through analytical tools. By definition this flux of being resists any form of reification and can be approached by intuition (unmediated cognition) only. The objects that we consciously identify as demarcated portions of reality are therefore secondary to this reality and the result of a reasoning process. Following this line of reasoning, for most of the time, dynamic objects in our environment are hidden. They are invisible and come into being, as a phenomenological event, only as we attend to them, and even then they come into being not as objects that come before our cognition, but as a mysterious occurrence that provides the possibility for cognition. Think about your socks. They definitely exist in a physical space, but most of the time you are not aware of their existence. As a phenomenological event, they come into being when they are brought forth into our consciousness, such as when you feel them to be too tight.

The argument is that objects do not simply exist waiting for our mind to heed them. In fact, as singularities and unique loci of our reality, those dynamic objects do not exist as individuated objects (having an identity as "a cat" or "a carrot") before we approach them as such. For example, a particular cat does not exist as "a cat" before I approach him through a general scheme that categorizes it as a specific member of the cat class. I do not deny the fact that something exists "out there" (where is this "out there"?), but just point at the fact that the something that exists out there exists as something as long as we approach it as something more general than a singular and yet indeterminate being which it really is. In fact, the singularity of the object is an ultimate barrier for its recognition as an "object." Since we interact with singularities (e.g., always with that being we reflect upon as this cat) and since by definition singularities are always unique events in the flux of being that cannot be fixed or demarcated by mental categories, we must reason that our basic interaction with the world is not with objects through reflection, but with dynamic objects through some unreflective correspondence.

We interact with primordial singularities in their flux of being and not with individual objects obeying the law of identity ($A=A$). When we attend to an object in our environment this object does not a priori exists as such. It comes into being primarily through the *intimacy* we establish with it, and only then are we able to "objectify" it by conceptualizing it and giving it a name. Therefore, in our daily experience, and at the most basic level of being-in-the-world, we are primarily involved in activities. We interact with the world and with primordial singularities in the world, and through our interactions with the world, we constitute the mind and the world.

Although our basic form of being-in-the-world is pre-objective and intimate, it seems that the *sine qua non* of living systems is their ability to detach from the concrete and particular experience/event/act/flux of being. This property is manifested by their semiotic activity, by their ability to signify and to constitute their existence through a signification process. The most basic activity of signification is the constitution of the boundary between the organism and its environment through a signification process that differentiates between self vs. non-self. This semiotic ability is also evident in classification behavior[12]: The tiger "knows" how to differentiate between its prey and the rest of its world although each prey is a particular being, and each interaction with prey is a singular interaction. In this sense, classification or conceptualization is primarily associated with the *activity* of *differentiating* a given universe of discourse as evident from the tendency of the organism to establish its autonomy by constituting the boundaries between itself (the self) and the rest of the world (non-self). Only at a higher level of abstraction, and only among human beings, this activity of differentiation involves the use of symbols, which are abstract entities fixed by the law of identity: A tiger is a tiger! That is, while I am using the sign "cat" as the name of a specific class of individuals, I am not in the realm of a singularity I am interacting with, neither in the realm of the classificatory process of differentiating cats from dogs, but at the most abstract realm of symbols (the name we give to those classificatory processes). As already noted by Merleau-Ponty: "For to name a thing is to tear oneself away from its individual and unique characteristics to see it as representative of an essence or a category" (1962, p. 176).

I have associated the idea of sign with the idea of concept. However, those terms should not be confused. A concept may be understood as a "principle of classification, something that can guide us in determining whether an entity belongs in a given class or does not" (The Cambridge Dictionary of Philosophy, 1995, p. 148). Although non-human organisms and even humans behave according to classificatory principles (as if they have concepts in their mind), the use of symbols involves a higher level of abstraction. In other words, human beings use symbols in a unique way that differentiates them from other living systems. This system of signs cannot be identified with a concept as a classificatory procedure or with classes/sets as a collection of things, but it is the second level through which we reflect upon the set/class and fix it as a stable entity, specifically by giving the classification process a name: a cat, a number, etc.

The idea that symbols are fixed and stable entities may be disputed by differentiating between *nouns* and *verbs*. Can we consider verbs as stable and fixed entities? The answer is definitely no. But neither are nouns stable and fixed entities. They just look like they are and therefore they are mistaken for corresponding to real objects. In addition, verbs always assume nouns that are the "object" of the activity. There is no "eating" without an "eater" and there is no "running" without a "runner." Therefore, the differentiation between nouns and verbs does not seem to hinder the argument I develop in this book.[13] Moreover, as suggested elsewhere (Bateson, 1991), our fascination with nouns and entities should be replaced by a fascination with processes, that do not depend on a preconceived set of objects. The shift to such a language has to be accompanied by a radical transformation in our conceptions (Bateson, 1991), and the current book aims to lead in this direction.

The differentiation between the symbolic and the non-symbolic form of existence can easily be demonstrated by turning to the non-human realm. For a mouse to meet a cat means to encounter a singularity (a dynamic object) which we, as reflecting subjects, consider to be a "member of the cat class." However, even the mouse does not meet the particular cat as a singularity in the sense that he judges him according to his specific and unique properties. This particular cat may be fond of mice, content, or simply a pacifist. However, a mouse that judges each cat it encounters individually as a singularity may find himself very quickly evolutionarily wrong. The mouse is a semiotic organism the same as all other living systems on earth. Therefore, it encounters the particular cat through a "class-concept," a member of a specific set in the same sense that I met "a man" today who is a class-concept of the class "men." However, the difference between the mouse and a human being is in their ability to move upward in the semiotic ladder. Although a mouse operates according to a classificatory process that motives it to avoid the company of cats, I doubt whether the mouse has the sign "cat" in his mind, at least in the sense of *a symbol that can be manipulated and communicated by his [its?] "sign community" without any reference to a concrete and immediate experience..* Although a mouse can warn his community of mice about an approaching cat, it is questionable whether the mice community will discuss this incidence at the evening dinner, culturally deliver this knowledge across generations, and improve their survival skills in the face of their mythological enemy. Being familiar with the politically correct culture of the current academic life, I must qualify the above argument by saying that this argument does not hold for Disney's hero – Mickey Mouse – and his lovely and highly intelligent spouse, Minnie.

In short, only human beings have the ability to reflect on their classificatory processes (which operate on their primordial experience with dynamic objects) by giving them names and fixing them into (what appears as) stable entities (signs) that can be exchanged beyond a particular context. In this sense, due to their relatively static and abstract appearance, signs may be used beyond a particular perceptually grounded context, and therefore *might be mistaken for having a referential meaning to some objects, whether on a Platonic or sensual realm.*

The operative nature of pre-semiotic, pre-objective and pre-reflective classificatory activity is sometimes difficult to conceive. This difficulty results from our bias to use semiotic systems for understanding more basic ways of being-in-the-world and from our attempt to use analytic tools for understanding our primordial being, which is the basic condition of mind rather than an object of the mind. For example, Set Theory is the most advanced formal semiotic system I know for handling a collection of objects. Within this system, a class/set/collection is usually defined according to its property (definition by extension). For example, "x is an even number greater than 3." In set theory notation: $A = \{x \mid x \text{ is an even number greater than } 3\}$. This statement defines the members of a certain class of objects. However, this way of considering sets may misguide us to believe that objects do exist and are simply represented in our mind.

Turning to the animal kingdom may clear this argument. The "cat" class can be logically characterized by a property of eating mice, a property that should definitely guide any rational mouse for a very specific behavior even if he is not familiar with set theory. However, a "classification process" can be described not only as concerning a collection of objects (e.g., cats) through some property (definition by extension) as common in set theory, but primarily as a *description of a relational structure*: If H is the relation "Hunt," "C" the concept of a cat and "M" the concept of a mouse, then C has the binary relational structure H to M: Cats hunt mice. Although we assume C and M as two objects (sets?) preceding the relation "H," this is meaningful only from our theoretical and reflective point of view. From a phenomenological perspective, we do not have to assume any familiarity of the organism with the abstract objects: cats and mice (and with their re-presentation in mind). We may assume that mice are familiar with the basic ideas of set theory and that somehow they *represent* sets in their minds. However, this hypothesis is either humorous or scientifically flawed. It is more reasonable to assume, especially from an evolutionary point of view, that the *relational structure* is the one that determines (at least for the mice) the meaning of M and C. *In this sense, mice live in a world of activities and tendencies to operate in certain contexts concerning certain singularities without assuming the re-presentation[14] of real objects in their mind. That is, the mice patterns of activities define both their identity and their phenomenological world. At the most basic level of existence, we are no different. However, we have the ability to suppress this basic level of existence through higher levels of abstraction.*

In our case, the idea is that at the pre-objective sense things may have meaning only as a mode of a *relational structure*, a basic mode of being-in-the-world, and not as objects that may be semantically defined and "re-presented" as static, canonical entities in the cognitive system. In other words, at the most basic level of analysis, we interact with singularities in a way we cannot reflect upon through our semiotic systems. We consider this primordial experience with singularities through classes. Those classes are not regarded in terms of *extension* (static property that characterizes the class), but in terms of *a process of differentiation and tendencies to operate in certain situations in a certain way*. It is only at a later and more abstract level of analysis that we reflect upon the processes that constitute our mind in semiotic terms that may be meaningfully defined from an extensional point of view. This suggestion points at the way our inquiry into the reification of the world should proceed.

Cat-logue 1
The Cheshire Cat and Descartes

The participants: A young university professor (Dr. N) writing a monograph concerning the dynamic of the mind, and his white cat named "Bamba."

Bamba: Hi, Dr. Neuman. Would you mind scratching my back?
Dr. N: Ohh..mmm... sorry Bamba, but I'm very busy right now.
Bamba: Really? What are you doing?
Dr. N: I'm trying to present some ideas concerning the processual nature of being and the reification of reality.
Bamba: Sounds impressive. What do you mean?
Dr. N: Listen, it's a little bit difficult to explain this theory even to human beings. . . but. . . how about participating in a thought experiment that would clarify my ideas?
Bamba: Sounds great! You should remember that my ancestors have contributed enormously to the development of psychology by participating, involuntarily of course, in lab experiments. In fact, my great-great-grandfather was the one who helped Tolman to develop his idea of cognitive maps. . .
Dr. N: Ok, ok, let's cut it. Now listen. Try to describe what exists in your world. If I may make it short, I can tell that you will find that things almost exclusively populate your world.
Bamba: What do you mean by "things"?
Dr. N: I mean objects.
Bamba: And what do you mean by objects?
Dr. N: Do you want a definition? The Concise Oxford Dictionary defines an object as "a material thing that can be seen or touched" or as "a person or a thing to which action is directed."
Bamba: Very concise. That's why they call it the Concise Oxford Dictionary. And how does it define a "thing"?
Dr. N: A thing is defined as "a material or a non-material entity, idea, action, etc., that is or may be thought about or perceived."
Bamba: Very, very concise!!! Our British friends are very concise. However, do you think that they read Wittgenstein?
Dr. N: Why are you asking?
Bamba: Because the act of lexically defining a "thing" through this "spaghetti of words" makes this "thing," which I don't yet understand, meaningless. Wittgenstein was one of those who pointed at the difficulty of clarifying the meaning of a word by a lexical definition ...
Dr. N: Ok, ok ... Let's leave it and let me try another way. Please describe to me what you have done today.

Bamba: That's easy! When I woke up this morning I went out to the garden, then I sensed a movement in my perceptual field. In any case I acted immediately, jumped and . . .

Dr. N: So you feasted on a bird again!

Bamba: What do you mean by "a bird"? As I already told you, I don't understand the meaning of a "thing" and "bird" sounds like a "thing." There was a movement in my perceptual field and then I jumped...

Dr. N: You vicious creature, don't try to reduce your moral responsibility for this crime by sophist arguments. Just answer the simple question: What was moving in your perceptual field?!

Bamba: As I already told you, I sensed a movement and then I acted. There is nothing behind it. By the way, it was tasty.

Dr. N: Aha! When you are saying "I" sensed, "I" acted, you are admitting at least the existence of one "thing," which is the subject of sensing and acting. This thing is the "I," it is your self: *Cogito ergo sum*!

Bamba: So you are using Descartes as your last defense line. Very impressive. He is a difficult opponent. However, Descartes' view of reality is just one possible scientific framework. Have you read "Alice's Adventures in the Wonderland"? Do you remember the Cheshire cat? This is *my* scientific framework and it does not have any "I"!

Dr. N: Ok, I give up. You are definitely not trapped in a reified universe. Let's have a break and I will scratch your back.

Bamba: My pleasure.

Chapter 4
Beyond the Curtain or into the Looking Glass

From what rests on the surface one is led into the depths.
Husserl

> Summary: Several philosophers strive for an unmediated understanding of the world. However, it seems that unmediated cognition is impossible since cognition is by definition mediated by signs. Therefore, inquiring into the reified universe is primarily an inquiry into our semiotic activity.

In the previous chapters, I suggested that we reflect on our world as almost exclusively populated by objects and pointed at the quantum leap between this appearance and the dynamic and indeterminate nature of reality. I use the expression "reflect on our world" because I do not deny the fact that our physical encounter with the world is organized into stable structures (perceptions), but only that the specific reified ontology (what there is in reality), which is somehow derived from this encounter, is wrong. In other words, from the fact that our physical encounter with the world is organized as stable structures, we can infer nothing about the structure of the world, rather only about the structure of our mind-world relations in a specific cultural setting. More precisely, on the structure of mind-reality as being reflected upon by people at a certain point and loci of human civilization. Our mind may be such a device that it turns the flux of sensual experience into stable structures (and they are not!). However, those are the structures of our mind-reality relationship, not "objective" structures that exist out there in reality as stable and fixed Platonic entities. In this sense, the world is not an object and it is not populated by objects. It is something else. As suggested by Merleau-Ponty: "The world is not an object such that I have in my possession the law of its making; it is the natural setting of, and the field for, all my thoughts and all my explicit perceptions" (1962, p. xi). This position emphasizes the idea that the world is the stage upon which the human drama unfolds and not a distinct and a separate "object" we possess through our mind. This idea brings us again to inquire into mind-world relations.

Husserl suggests that objects in nature are known through perception, but *acts of consciousness* through reflection. In this sense, inquiry into our conscious experience is not an inquiry into reality per se, as if we are removing the curtain of appearances and confronting what Kant describes as the "thing-in-itself," but a journey into the "looking glass" that reflects our mind. Similar to Alice's adventures in "Through the Looking Glass," our journey to the far territories of the reified universe is in fact a journey into our mind, and from our mind back into the reified universe in a spiral-like process.

If we accept the idea that acts of consciousness are known through reflection, and that our reified universe is the result of a certain reflection on our most basic bodily encounter with reality, how is it possible to transcend our reflection in order to uncover the origins of our mind and to encounter Being in its virginity? Well, one possible suggestion on how to conduct this inquiry appears in Husserl's phenomenology.

In Chapter 1, I asked you to participate in a thought experiment that follows the phenomenological theory of Edmund Husserl, or at least some variation of it. Whether we accept Husserl's unique method of reflection, known as "phenomenological reflection," or not, the idea that the content of consciousness is available only through reflection is a key to the understanding of our reified conception of the universe. I want to discuss this point further, specifically by considering Husserl's philosophical theory.

Edmund Husserl (1859-1938) is considered one of the most influential thinkers in Continental philosophy. He sought to develop a descriptive science of consciousness, which is grounded in the unmediated examination of phenomena or appearances. According to this method of inquiry, explanations, hypotheses, theories, predisposition or any other prior structure should not be imposed on the phenomenon under inquiry before the phenomenon has been understood from within, as a pure and virgin object given to the reflecting subject, as an intuitive and insightful truth similar to the one the subject experiences when he reflects on the arithmetic expression "$2 + 2 = 4$."

The attempt to achieve purity is by no means a novel venture. The Platonic search for abstract ideas or the Cartesian venture of Rene Descartes are just two famous examples of philosophical endeavors that aim to get rid of misleading factors in order to uncover the latent structure of reality. As we recall, in his seminal essay, "Meditations," Descartes sought to establish the architecture of knowledge by grounding it on firm foundations. In order to achieve his pretentious aim he used a method of tactical skepticism in which he doubts the existence of everything, including his own existence. This method led him to the idea that even at the extreme phase of skepticism, of questioning his own existence, there is someone or something which is the agent of the skeptical activity. Hence, the most basic and unshakable source of our knowledge is the mental self, evident through the thinking activity: I think therefore I exist, which sounds more impressive in Latin: *Cogito ergo sum*. From this position, known as Solipsism or transcendental subjectivism, from the undeniable existence of the mental ego, Descartes proceeds to prove the existence of God and the existence of the material world, the world of objects. Husserl acknowledges his intellectual debt to Descartes, or more particularly to Descartes' radical approach to philosophy, by naming one of his essays "Cartesian Meditations." However, Husserl's phenomenology differs substantially from the method presented by Descartes in several important senses, such as by *taking the world of appearances* (phenomena) as its point of departure.

Phenomenology seeks the description and structural analysis of consciousness as it is experienced without the mediating force of other systems. This is Husserl's *first methodological principle*. In order to achieve this aim Husserl offers the method of "phenomenological reduction" (*epoche*) in which the subject transcends the idea that there is a world independent of experience and focuses on the intentional act of consciousness. This method takes the world of phenomena, of appearances, whether real or imagined, as its starting point. Then, and through the method of phenomenological reductionism, all objects become reducible to their purity: "The epoche can also be said to be the radical and universal method by which I apprehend myself purely" ("Cartesian Meditations," 1973, p. 21). This extreme striving for purity seems to be an impossible reductionist venture, since purity, as we know, exists in the mind of certain philosophers and clergymen alike, and only there. Indeed, one of the main figures in the phenomenological movement – Merleau-Ponty – argues that the most important lesson that the phenomenological reduction teaches us is the *impossibility* of complete reduction! This impossibility teaches us an important lesson concerning the mediated nature of our mind and even provides us with a hint about how to transcend this mediation and to encounter the primordial and pre-objective experience in its purity. This idea will be illustrated later concerning the idea that a *process of differentiation* is the purest and most basic result of our phenomenological reduction. For the current phase of our inquiry, we should just recognize the interesting direction toward which the phenomenological venture points.

Husserl's system is also relevant to our inquiry because it transcends the common division between empiricists and rationalists: those who believe that the source of knowledge is the sensual data and those who believe in innate forms of mind we impose on the outside world.[15] If we adopt the phenomenology of Husserl, then we cannot accept the classical empiricist position that traces the source of knowledge to our sensual experience, because we know that our experience is mediated by folk conceptions, daily experience with the world and other sources of distortion (mediation). On the other hand, we cannot accept the idealist or classical realist position that knowledge exists independently of our mind, because all objects are encountered perceptively, that is, from a specific perspective of the individual with his unique properties. This critique has been summarized by Merleau-Ponty as follows: "Empiricism cannot see that we need to know what we are looking for, otherwise we would not be looking for it, and intellectualism fails to see that we need to be ignorant of what we are looking for, or equally again we should not be searching" (p. 28).

Husserl's desire for purity aims to achieve an unmediated and pure view into reality. Whether his theory is philosophically significant is a contestable issue, like many other issues in philosophy. However, the attempt to encounter reality through "intuition" or unmediated activity is something that deserves profound discussion. One of those who took this challenge was the famous American polymath, Charles Sanders Peirce.

Peirce opens one of his most famous essays (1978) by asking whether we can judge whether cognition has been determined by previous cognition or by intuition. By intuition Peirce means "a cognition not determined by a previous cognition of the same object, and therefore so determined by something out of consciousness" (p. 135). The question is important since, if we argue, like Husserl, that intuition is possible, then we should be able to differentiate (or to provide clear criteria) between cognition which is mediated and cognition which refers immediately to its object.[16] The issue is even far more complex than it seems, because if we do not have the possibility to cogitate "intuitively," as suggested by Husserl, then one must supply an answer to the question: How is it possible for a subject to transcend its own boundaries and to use abstract entities (i.e., concepts) that are not grounded directly in his environment? If we do not have the ability to approach the essence of things intuitively, how is it possible to have abstract concepts that cannot be traced to our perceptual experience? If cognition is determined by previous cognition, we should assume a chain of cognitions that stretches from certain dynamic objects in reality to the most arbitrary symbols that populate our minds. From the most concrete cars in our environment to the abstract concept "car" and for the unique symbol we use to name this concept. This is the *symbol-grounding problem* (Harnard, 1990). It is the six million dollar (or more) question of epistemology, and I am not sure that Peirce solves it by denying the possibility of intuition. Nevertheless, let us proceed and examine Peirce's arguments against the idea of intuitive knowledge.

Peirce shows that there are many instances in which we are sure that we conceive an object directly, without any form of mediation, while the state of affairs is totally different. For example, when we look at a painting we may conceive it as a three-dimensional entity with a certain depth. However, the painting is clearly two-dimensional and even if it were a three-dimensional entity, the human eye (or mind) cannot be considered a three-dimensional space or gallery for the presentation or transformation of three-dimensional objects. So, how do we see an object as a three-dimensional object?

The idea is not to open a discussion concerning the constructive power of our mind to create unique representations out of the blue, but to show that there are many instances in which we consider a certain object as intuitively available while the object is the result of a previous cognition/s or a mediated activity that constructs it. This is a strong argument against the idea that we can differentiate between intuition and mediated cognition. However, this argument is theoretically flawed in other senses and Peirce's most convincing attack against intuitive knowledge is supported by his semiotic[17] theory. In this context, we should re-frame the issue of intuition vs. mediated cognition by the question: *Can we think without signs?* But what is a sign? And why are signs relevant to our discussion?

Peirce defines a sign (or what he calls a "representamen") as "something which stands to somebody for something in some respect or capacity." A sign is always something that stands for something else for a referent. It should not be identified with the object it signifies, as a map should not be confused with a territory, and a menu with food served. Taste a menu in a restaurant and you will notice immediately that the difference between the sign (signifier) and the referent (signified) has a solid base. This example definitely qualifies for restaurants where the taste of the menu and the taste of the food differ. We can graphically illustrate Peirce's conception of the sign as follows, where the triangle does not represent a stable structure, but a *process* of signification:

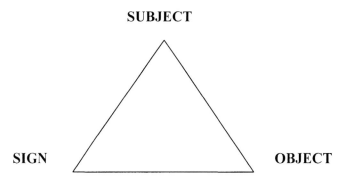

Figure 1. Peirce conception of the sign.

The idea that we approach the world through signs and that signification is a process rather than a static representation is especially relevant for our analysis. To recall, we reflect on our world as populated by objects. However, if our access to reality is mediated by signs that are themselves not fixed entities corresponding to "real" objects, but a position in a wider process of signification, how do we know that objects really populate reality and that they are not some form our mind imposed on reality through its mediated semiotic activity? The answer to this important question is that, indeed, there is no cognition without signification and that objects do not exist in reality, but emerge out of mind-reality through a process of signification. Meanwhile, the reader is invited to ponder this idea while we proceed in our inquiry.

Peirce differentiates among three types of signs: icons, indexes and symbols. An *icon* is a sign which is perceptually similar to its referent, the object to which it refers. For example, a sign of a cow over the butcher's shop clearly resembles the creature from which the butcher makes his living.

An *index* is a sign which has a correlation with its referent. When a chimpanzee wakes his group, warning them through sounds that there is a predator in the area, the sounds do not have any resemblance to the approaching enemy, but a correlation only.

A *symbol* is the most abstract (or decontextualized) type of sign since it lacks physical resemblance or correlation to the referent. When I asked you what exists in your world and you replied with an answer that includes things like "a chair," you actually reflected on your consciousness through the use of a sign whose meaning seems to be derived from a social norm per se. Although the sign "chair" may have certain perceptual stability across contexts, the sign itself is an *abstract* one in the sense that it does not have either perceptual resemblance or correlation to the thing it signifies. This is a property the famous semiotican Saussure described as the *"arbitrariness"* of the sign. However, it is more than that. When you use the sign "chair," you are using a sign which is abstract in the sense that it does not refer to a specific object in a specific context. You may describe a specific object like "This chair" or "That chair." However, the concept of "chair" seems to exist above the particulars it represents and this "abstract" nature is what allows it to be used freely across different contexts. This is an amazing property of the symbol that turns it into one of the most powerful tools of human intelligence. When I come to a friend and he says to me: "Please sit on the chair," I know exactly what he means, although the chair in his house may be perceptually different from the chair in my house. When I ask my older daughter how many marbles are in her hand and she replies "2," and when I ask her how many rings are in her hand she also replies "2," she uses the same sign – 2 – to signify two different kind of things that do not have the slightest perceptual resemblance to their references, and it works. This looks like magic and, indeed, in many cultures, from antiquity to the present, the power of signs has been considered a sacred thing that originated from the Deity himself.

The meaning of a symbol seems to be determined only by convention. Many words we use in order to communicate are symbols, although they can be historically traced to somatic experience or to icons (Lakoff & Johnson, 1999; Sebeok & Danesi, 2001). This state of affairs presents us with a mystery that bothers semioticans and children alike. Anyone who has children has encountered the general question: "Daddy why do we call X 'X'?" For example, "Daddy, why do we call a cat a 'cat'?" As parents, we answer this question by saying something like: "I don't know" or "This is the way it is." This answer is clearly not a satisfactory answer. Why is it the way it is? Why do English speakers call a cat a "cat" and not a "gato," like Spanish speakers do?

Peirce's answer points to the conventional nature of symbols, but not to the origins or the dynamic of these conventions. This is an important point, since Peirce is unable to cope with the symbol-grounding problem and to explain how a shared somatic encounter with reality, which is located in a social process of signification, diverges into many symbols: All people see a cat as a cat, although the number of symbols we use to denote this creature enormously varies across social groups and time intervals. Later, I will try to provide my own answer to this mystery. I must warn the reader that I do not have an answer to the specific question why do we call a cat a "cat" and not a "gato." However, I have some ideas regarding the dynamic of signs and the reason why the name we attribute to a classificatory process cannot be easily traced to its somatic origins. For the current inquiry, the reader should think about the idea of chaotic systems, in which infinitely small differences in the initial state of a deterministic and non-linear system lead to unpredictable, diverging trajectories as time unfolds. If this kind of dynamic characterizes the sign system, then it is possible to explain the way the same perception anchored in the same experience of people with the same bodily structure diverges into so many different symbols.

Now let us return to the question: Can we think without signs? For Peirce the answer is clear: All thought must necessarily be in signs. Since thought is always *about something* (the intentionality discussed by Brentano) and not the something in itself, we (and other living systems as well) cannot think without signs. The world of phenomena comes into being only through the unique form of beings who express the ability to detach themselves from reality and to reflect on their experience through the use of signs. As suggested by Merleau-Ponty:

> As soon as there is consciousness, and in order that there may be consciousness, there must be something to be conscious of, an intentional object, and consciousness can move towards the object only to the extent that it "derealizes" itself and throws itself into it, only if it is wholly in reference to ... something, only if it is a pure meaning-giving act. (1962, p. 121)

In sum, Peirce's claim, provided by the insights of Merleau-Ponty, is that all thought takes "the form of overt linguistic activity, conscious inner dialogue, or unconscious inference which is modeled on linguistic behavior"[18] (Hookway, 1985, p. 27). There are no direct perceptions through intuition (unmediated thought). Everything that we are conscious of are signs resulting from inference and "man can think only by means of words or other external symbols" (Hookway, 1985, p. 188).

From this analysis, we may conclude that since signs mediate thought, any serious inquiry into the origins of our reified universe must involve a *semiotic* perspective that takes into account the fact we are sign-using creatures. This statement does not exclude the fact that other living systems are sign users and that the semiotic perspective should be carefully applied in order to understand them too. However, human beings are a special kind of sign-using creature and this fact should not be masked by similarities with other creatures. Moreover, it seems that there is a way to reconcile between Husserl and Peirce. If we inquire deeply into our sign system, we may find ourselves reflecting on the origins of our mind, and transcending the boundaries of semiotic mediation. In this sense, Husserl's struggle for purity may be reconciled with Peirce's semiotic stance.

We opened this book by questioning the reified conception of reality and by arguing that our basic encounter with reality is interactional and objectless. The question that inevitably results from this analysis and from the need for a semiotic reflection is: Why is it that the structural stability of our perception turns to indicate the structure of reality? Are people blind to the fact that the mind is the one that actively organizes their somatic experience through the mediating power of signs, and that it cannot reflect the structure of reality as it "is"? How is it that people force on the world the same structure of their mind? Is there a semiotic interpretative framework that can make sense out of this mystery? The next chapters discuss several semiotic issues that may advance our understanding of the reified universe.

Chapter 5
A Guided Tour in the Kingdom of Signs

Protos Logos (At the beginning was the word)
John 1:1

> Summary: This chapter presents two main conceptions of the sign: non-developmental and developmental. The ability to use signs is described in the Bible as a unique property of the human mind. This position does not confront developmental aspects of semiotic systems. Evolutionary theory confronts the symbol-grounding problem by locating the origins of our symbol system in simple communication systems of lower ordered organisms, and by explaining the emergence of the symbol system through Darwinian mechanisms. This perspective, too, fails to explain the emergence of the sign. This chapter concludes by pointing at the importance of adopting a social perspective for inquiry into semiotic systems.

Let us summarize the arguments presented so far and point to the next steps of our inquiry. First, we discovered that we reflect upon reality as reified, a reality populated by objects. In order to trace the origins of this reflection we turned to inquire into our consciousness. Following Peirce, we realized that such an inquiry should begin with the assumption that our mind is mediated by symbols and that we cannot approach objects, whatever those entities are, in a directly intuitive way. This position does not overlook the possibility that direct confrontation with the pre- or the trans-objective is possible. It just points at the need to analyze carefully our sign system as a springboard for approaching the primordial origins of mind.

In general, we can differentiate between two main perspectives of semiotics: the non-developmental and the developmental. We can call the first perspective "The Myth of the Fall." This perspective suggests that human symbolic activity is complete and that there is no need for developmental analysis in order to understand it. This perspective is traditionally associated with the idea that there is an essential link between the sign and that which is signified. The language of magic and witchcraft depicts this idea by considering signs as essentially (rather than arbitrarily) coupled with their referents. "Let there be light," said the Almighty and light there was. "Horses," said the good fairy, pointing at the mice, and the amazed Cinderella saw the mice turn into horses. In both of those cases, the symbols are closely associated with their referents. They generate them or represent their essence. The second perspective, entitled "The Myth of Progress," suggests that our unique semiotic system is the result of a long historical evolution. In this context, we may ask whether there has been some progression in the semiotic ability of human beings and that our ability to manipulate abstract-arbitrary signs (symbols) has evolved out of iconicity (or even from a non- or pre-semiotic phase), whether through a long period of biological Darwinian evolution,[19] through a shorter period of cultural evolution, or whether evolutionary explanations are meaningless and the ability of human beings to manipulate symbols is a fact that should be studied without pondering any developmental issues. Let us keep those questions in mind and start our journey in the kingdom of signs.

5.1. THE MYTH OF THE FALL

Today linguistics is mainly associated with the name of Noam Chomsky and his a-historic and abstract mathematical theory of language. By presenting and supporting the idea that universal structures govern our syntax, Chomsky has reached the position of cultural icon and his dominance is felt everywhere, even in fields such as politics, in which his expertise has no advantage over the layman. We should keep in mind that Chomsky's abstract, mathematical, and a-historical theory of language (or more accurately of syntax) is in sharp contrast with the traditional linguistic research that governed European linguistic philological research for years. European linguistic research not only studied the history of languages, but was *obsessed* by the origin of languages, the uncovering of their relationships, and the reconstruction of the lost proto-languages from which families of extant languages descend (Sampson, 1980). The dominant conception was that by studying language/s in the same way we study geology or botany, a new science of language can be established on par with the natural sciences. This idea was clearly expressed by Max Müller, one of the great proponents of this venture:

> ...the language which we speak, and the languages that are and that have been spoken in every part of our globe since the first dawn of human life and human thought, supply materials capable of scientific treatment. We can collect them, we can classify them, we can reduce them to their constituent elements, and deduce from them some of the laws that determine their origin, govern their growth, necessitate their decay; we can treat them, in fact, in exactly the same spirit in which the geologist treats his stones and petrifactions, - nay, in some respects, in the same spirit in which the astronomer treats the stars of heaven, or the botanist the flowers of the field. (Müller quoted in Harris, 1987, p. 7)

The above statement may cause us a momentary feeling of intellectual perplexity. Is this the man who wrote the "Origin of Species"? Or is he the spiritual linguistic twin of Charles Darwin? Indeed, there are some very interesting similarities between the linguistic venture of the nineteenth century, with its quest into the origin of languages, and the evolutionary theory of Charles Darwin, who sought the origins of species.

There are, of course, some substantial differences between the two ventures. For example, the Darwinian approach to evolution, at least concerning language, suggests some kind of progression from primitive communication systems to human language. In contrast, the theological spirit of the European linguistic philological research adheres to the myth of the fall and to the idea that human beings came into the world with a complete semiotic capacity. Since Adam and Eve spoke in a language that communicated with God, our current languages cannot be considered as a sort of progress. Therefore, the linguistic research of the nineteenth century was deeply rooted in the same theological conceptions that Darwin's theory challenged.

It is worth discussing classical European linguistic philological research because it was occupied, in a similar way to Darwin's theory of evolution, by the issue of *origins*. For example, one of the most central questions that occupied linguistic research of the nineteenth century was: Which language did Adam and Eve use? Which language was the language of Eden? Did Adam and Eve speak in Hebrew – the holy language of the Bible? Did they speak Sanskrit in the Garden of Eden, which was located somewhere near the Ganges?

Those questions may seem nonsense to the modern mind since the biblical story is not considered by the current secular scientific establishment to be a legitimate scientific source of evidence. However, for European scholars occupied with theological and ethnic issues, the question of which language Adam and Eve used to speak with God was of prime importance. After all, uncovering that language may reopen the impaired channel of communication with the Almighty!

To the best of my knowledge, Christian philologists dealing with the origins of language were not familiar with a rich Talmudic tradition of commentary on the Book of Genesis and the origins of language. This homiletic interpretation of the Scriptures is called *Midrash*. Genesis Rabba is the first complete and systematic Judaic commentary to the Book of Genesis, which was redacted at A.D. 400. What did the Talmudic sages think about the role of language in the Book of Genesis? Let us read the following excerpt:

> A. R. Berekhia in the name of R. Judah b. R. Simon commenced [discourse by citing the following verse], "By the word of the Lord were the heaven made, and all the host of them by the breath of his mouth" (Ps. 33:6).
> B. Not by hard work or by toil, only a word.
> (Genesis Rabba, p. 28)

This excerpt points to an interesting interpretation of the sign, some of my modern readers (and I doubt whether there are "modern" readers) may have forgotten. The act of producing a word (or a sign) is not an act of signification, and God does not signify things. The word is the *event* of creation, and therefore Creation, in its various modus, is immersed in the sacred power of the Lord as made material by his words. Please note that God does not seem to believe in hard work ("Not by hard work...") but only in the power of words.[20] In this sense, God is clearly not a Protestant ...

As long as this process of creation concerns God, the symbol-grounding problem makes no sense. God is not a signifying being, but a creating deity. As concerning men, things are quite different. Read the following excerpt taken from Genesis Rabba:

> A. Said R. Aha, When the Holy One, blessed be he, came to create the first man, he took counsel with the ministering angels. He said to them, "Shall we make man?" (Gen. 1:26).
> B. They said to him, "what is his character?"
> C. He said to them, "his wisdom will be greater than yours."
> D. What did the Holy One, blessed be he, do [in order to make his point]? He brought before them domesticated beasts, wild beasts, and fowl. He said to them, "As to this creature, what is its name?" But they did not know.
> E. "What is its name?" But they did not know.
> F. Then he brought them before man. He said to him, "As to this, what is its name?" "Ox." "And as to this, what is its name?" "Camel." "And as to this what is its name?" "An ass." "And as to this what is its name?" "Horse."
> G. That is in line with this verse: "And whatever man called every living creature, that was its name" (Gen. 2:19).
> H. He said to him, "and what is your name?"
> I. He said to him, "As for me, what is proper is to call me 'Adam,' for I have been created from the earth [which in Hebrew is adama]."
> J. "And as for me, what is my name?" He said to him, "As for you, it is fitting for you to be called, 'The Lord,' for you are the Lord of all that you have been created." (Genesis Rabba, p.183)

The first interesting thing about this excerpt is that the Talmudic scholars describe God as a democratic governor who councils with his angels before making a decision. The issue he is discussing with the angels is whether to create Man. A crucial issue which is relevant for all of us. At least to those who consider themselves human. The counseling angels try to take their job seriously by raising some questions concerning the nature of the planned being. They know their boss and that good advice pays. In order to decide rationally whether to support the new product, the angels question its character. It is strange that they seem to know what a Man is, and question his character, although they were told nothing about him. It is as if they are saying: "Hey, we don't care what the nature of this man you would like to create is, just don't make another nasty creature with a bad character like the alien you created yesterday."

God's answer to the angels' query is interesting: "He said to them, 'his wisdom will be greater than yours.' In order to prove this point, God brings the other creature (which according to the story has not been created yet!) and asks him to *name* the creatures God created. Therefore, the first exam in the universe seems to involve a semiotic task concerning certain entities (well-demarcated portions of reality: beasts and fowl) that precede the semiotic process. God does not only test the ability of his creature to name things, but also asks him to explain the *reasons* for the names he gave.

The explanation Man provides to God's questions shows that the names (the signs) he uses *reflect the essence of the signified object*. For example, he names himself "Adam," since he was made out of "adama" (earth in Hebrew). This name associates the sign with the essence of that which is signified, and points at the simple referential meaning of the sign.

From a psycholinguistic point of view, Adam is even far more intelligent than that because he names God the "Lord," according to his *relation* to the world. Ask a young child to name an object and to explain his activity. You will see that relational explanation is an advanced developmental phase. Therefore, our story ends when Man successfully passed the same exam the angels failed. What this story tells us is that (1) wisdom is associated with the ability to name (to signify) objects that precede the signification process, (2) signs signify the essence of those things, (3) the social is absent from the signification process, and (4) *Man's semiotic faculty is complete* and includes the symbolic aspect. At least according to the above story, the symbol-grounding problem is not a problem since symbolic capacity was built by God into the human mind.

I am not sure whether God read John Austin's "How to Do Things with Words," but I assume that Austin did not read Genesis Rabba. While the power to do things with words, in the sense of creation, is the unique property of the divine creator, human beings have the unique ability to name the things created by God to signify objects that exist in reality, and to communicate those signs. That is, human beings have the semiotic ability to *represent* and to *communicate* these representations fully and by nature, at least according to some conceptions presented in Genesis Rabba. The above Talmudic story could have been accepted as a legitimate starting point for a scholarly discussion: Semiotic ability may be considered as a unique faculty of man and one should study it as a given *system* of representation and communication without getting into the mystery of its origins. This is the systemic theory of semiotic presented by Saussure. However, there are other conceptions that consider our semiotic ability through a developmental perspective. The most famous of these approaches is the Darwinian evolutionary approach that considers human language as a phase in a long evolutionary development. The next section aims to inquire into this perspective.

5.2. THE MYTH OF PROGRESS

The idea that the ancestors of humankind were non-linguistic beings did not even enter the mind of those scholars who dealt with the language of Eden. After all, the Book of Genesis as undisputed scientific evidence tells us that Adam and Eve were the first of their kind and that they gave names to the creatures God created. If they were able to name those creatures and to speak with God himself, then the existence of language for the royal couple of humankind seems to be an indisputable fact.

From a Darwinian perspective, things may look quite different. Darwinian theory argues that human beings are not the descendants of the royal couple from Eden, but rather the descendants of simpler and qualitatively different forms of life. Indeed, one of the major ideas in Darwin's theory is that the different species that now populate the earth have common ancestors (Sober, 1993). According to this theory, human beings have evolved out of simpler forms mainly through the mechanisms of natural selection. If we adopt this general perspective on the origin of life for human language, then the inevitable conclusion is that our ancestors were not complete linguistic beings as the Bible tells us, and that human language was not given to Adam and Eve ex nihilo (or ex Deus), but evolved out of simpler systems (e.g., primitive signaling systems) of communication. Let us dwell a little bit more on the evolutionary theory in general and the evolutionary theory of language in particular.

There are three important components of the Darwinian theory of evolution: variation, heredity and differential fitness (Brandon, 1996). Significant diversity in morphological, physiological and behavioral traits among members of species is considered variation. Heredity is the process through which the traits just described are heritable from parents to offspring. Differential fitness is the fact that different types of organisms leave different numbers of offspring in the short- or long-term, and, today, heredity is explained by genetics.

The mechanism that stands at the heart of Darwin's theory is natural selection. Darwin considers it the ultimate explanation for differential fitness. In this sense, fit (or adapted) organisms are those that survived the constraints imposed by the environment as evident from their offspring. This idea is expressed by the following definition of relative adaptedness:

> a is better adapted than b in E [its ecological niche] if and only if a is better able to survive and reproduce in E than is b. (Brandon, 1996, p. 23)

The idea is that traits in a given organism (e.g., symbolic ability) are functionally associated with its evolutionary history. In other words, each trait of an organism can potentially be explained by pointing at its evolutionary function or to its contribution to the adaptation of the organism. This thesis has been adopted for explaining the evolution of natural language through natural selection (Pinker & Bloom, 1990), and it is sometimes accompanied by sophisticated mathematical models (e.g., Nowak & Krakauer, 1999). However, while the natural selection thesis may be supported[21] with certain evidence concerning the development of physiological organs such as the eye, things are far more complicated when considering human language.

In order to point at the difficulties of using a Darwinian evolutionary explanation for the phenomenon of natural language in general and the symbol grounding problem in particular, let me present one specific problem – the continuity discontinuity debate (Aitchison, 1998) – that occupies evolutionary biology, and its relevance for the understanding of language.

Communication among animals covers a very specific and limited number of activities such as telling others that there is food and where it is, alarm/warning, territorial claims, recognition and greetings, indication of emotional stress, and so on. Think about the communicational behavior of a dog or a cat and you will notice that this behavior is extremely limited when compared to communication among human beings. On the other hand, communication between animals is far more complicated than was considered in the past. Although communication among animals has been usually described as a very simple activity, it has been discovered lately that "higher order organisms" (e.g., chimpanzees) express complicated communication activity, such as the one evident from their deceptive behavior. This is evident from their ability to manipulate their community members in order to achieve some personal material benefits.

For example, Griffin (1981), who deals with the question of animal awareness, mentions an incident in which a female wolf was trying to consume a piece of meat while being bothered by her cubs. The wolf deceived her cubs by making a sign of alarm when there was no real danger evident in the environment. After the young wolves escaped to their cave, the female wolf could finally enjoy the meat without sharing it with demanding offspring. This example shows that non-human organisms are far more complicated than have usually been portrayed. However, here comes the difference. Although higher order animals are able to use signs without a concrete reference in reality, the ability to give *names* to things and to use those symbols beyond a particular *operational* context is an exclusively human ability. Even if chimps, like the famous Washoe, learn several words arbitrarily associated with physical objects, they do not have the ability to use them freely across contexts or to string them together into a well-structured sentence. In this context, the term "design feature" may be quite helpful to sharpen the difference between the human and the non-human semiotic systems.

Hockett (1960) uses the term "design feature" in order to indicate a property which is present in some communication systems and not in others. He lists sixteen design features of human language. For example, *displacement* is a design feature which describes the fact that language can be used to talk about things that are remote in time and place from the interlocutors. For example, "I saw Danny yesterday at your parents' house." Another design feature is *arbitrariness*: there is no iconicity or physical resemblance between a linguistic sign and the elements of the world to which it is connected. This feature has been previously mentioned. Those differences bring us back to the Garden of Eden because the linguistic behavior which is most evident among its human residents is the symbolic one (naming). Only Adam and Eve express the ability to name things and to construct utterances as they speak with God... with one anomaly of course. The mythological serpent of Eden expresses a very high linguistic proficiency. In fact, a close analysis of his dialogue with Eve shows amazing resemblance to the rhetoric used by modern salespersons (Nash, 1992), a finding that may suggest that the idea of incarnation may not be so unreasonable after all!

At this point, we may recall the question that opened our current discussion. If we are evolutionary linguists, we must be determined to trace the way our linguistic ability has evolved out of primitive forms of communication. Has it developed through a continuous and smooth process or a discontinuous process? What is important to note about the difference between human beings and primates is that simple semiotic activity, which is evident in simpler organisms, is a necessary condition for naming ability evident in human beings. In this sense the shift from simple semiotic activity to naming may be described as a *continuous* process. On the other hand, naming activity involves a qualitative leap, which is not predictable from simpler symbolic activity and therefore represents a *discontinuous* line of development. In other words, naming may be described as both a continuous and a discontinuous process at the same time! More complicated than we thought.

The important thing about the evolutionary theory of language is that it tries to locate language within a wider context of our biological evolution. Unfortunately, this popular venture suffers from severe theoretical difficulties, and it cannot explain the development of symbolic activity and its development among human beings. Indeed, many sophisticated evolutionary models of language take the arbitrariness of the sign as their starting point and not as a crucial feature that should be explained.

Chomsky is sharply critical of Darwinian evolutionary models of language; he considers language as a unique device that cannot be accounted for by the Darwinian evolutionary perspective. In his book "Language and Problems of Knowledge" (1988), Chomsky expresses his general dissatisfaction with Darwinian evolutionary theory in a clear and convincing way:

> The outstanding American philosopher Charles Sanders Peirce, who presented an account of science construction in terms similar to those just outlined, argued in this vein. His point was that through an ordinary process of natural selection our mental capacities evolved so as to be able to deal with the problems that arise in the world of experience. But this argument is not compelling. It is possible to imagine that chimpanzees have an innate fear of snakes because those who lack this genetic determined property did not survive to reproduce, but one can hardly argue

that humans have the capacity to discover quantum theory for similar reasons. The experience that shaped the course of evolution offers no hint of the problem to be faced in the sciences, and ability to solve these problems could hardly have been a factor in evolution. (p. 158)

Beyond the lack of scientific evidence, that clearly supports this evolutionary theory of language, there is another disturbing issue that concerns the inappropriateness of using a biological explanation in order to explain a phenomenon which is clearly non-biological in the sense that its subject matter is not a biological entity. Let me try to illustrate this difficulty by using a simple example. Suppose that a friend invites me to dinner. A couple of minutes after the dinner starts, I address my host and say: "Ori, could you please pass me that excellent salad you made? I cannot resist having another helping." This is, of course, a very common interchange. By using the force of language, I politely asked my host to pass me the salad. In other words, when I ask my host, "Ori, could you pass me the excellent salad you made," I am not questioning his physical ability to move the salad from one point on the table to another, but am asking him, in an indirect way, to deliver the salad to me. I could command him: "Bring me the salad!" or just ask him to do it: "Please pass me the salad." However, the linguistic form I use can be understood only through a *social* perspective of politeness.

In the above example, I also justify my request by pointing out the excellent taste of the salad and, again, justification is an argumentative move embedded in our social life. Although the above interchange may be reduced to the social aspect of human beings, and the social aspect may be explained by the Darwinian evolutionary mechanism, this kind of explanation seems to be too far away from an elegant scientific explanation. Can the Darwinian evolutionary mechanism account for the illocutionary force of language? Does variation and selection account for the use of argumentative moves (e.g., justification) during a conversation?

In principle, the answer is positive. Evolutionary theory is the possible explanation for almost every human phenomenon I am familiar with, from the sexual preference of women across the globe for powerful men, to the mysterious phenomenon of consciousness. However, those explanations seem to suffer primarily from a problem of "misplacement" that characterizes some of the most pretentious reductionist ventures of the sciences. By "misplaced explanation" I mean an attempt to use a simple explanatory device from one dimension of human existence in order to explain a phenomenon that exists on an orthogonal dimension of human existence. For example, physics is a highly respected scientific domain with enormous success in explaining the material world in which we live. No one is stupid enough to deny the fact that our bodies are constrained by the laws of physics. Try to walk through a wall or to push your thumb into a solid stone (as did the famous Dr. Johnson in order to prove the existence of reality to the Bishop Berkley) in order to convince yourself in this fact. However, trying to explain the speech event portrayed above through the laws of Newton is as senseless an attempt as trying to explain the movement of celestial bodies according to some social norms of behavior during dinner conversation. Does planet earth hold us tight to its surface because it is polite enough not to throw us into empty space? Or is there a physical law that can better explain this experience? From this analysis we must conclude that the social as a distinct realm of analysis plays a key role in understanding semiotic systems and that it should be carefully applied in our own inquiry.

Whether Darwinian evolution is able to explain the phenomenon of language also involves questions regarding the meaning of *explanation* and *understanding*. As usually happens in theoretical discussions, concepts are freely used without acknowledging that they have meaning only within a wider context or interpretative framework. The Latin etymology of the term "explanation" may be used as a springboard for discussing the difference between explanation and understanding. "Explanation" is composed of "ex" and "planus," which means that something is being presented outside on a plane. The best way to understand this activity is by imagining something that was hidden and now is explicit (ex plain) and visible on the simplest surface, a surface that cannot hide anything, a plane. Therefore, explanation involves an activity in which something hidden, the latent structure of reality, is uncovered, becomes visible and clear to any person who has eyes in his head and a mind in his skull (or out of it).

Wilhelm Dilthey, who attempted to lay the foundations for the humanities, makes the difference between explanation and understanding a cornerstone of his method. According to Dilthey, the natural sciences seek a causal explanation for natural phenomena or more accurately for their abstract representations. An explanation may be interpreted as an attempt to uncover the latent and abstract structure of a phenomenon by ignoring its concrete appearance and by formulating a model that represents the causal relations among its elements. This representation has a social aspect. The representation should be so lucid that the collective cannot deny its validity or at least that there is a consensus among a community of scholars that it is a valid representation.

In contrast with explanation, Dilthey used the term *"understanding" (Verstehen)* as a totally different approach to man and his world. "When we 'understand' something, we see it in its proper interrelationship of whole and part in the general swirl of mental life and experience" (Betanzos in Dilthey, 1988, p. 23).[22] Understanding, according to Dilthey, is clearly the same as intuition in Spinoza's philosophy, and both point at the unique activity of approaching a phenomenon holistically.

Following Dilthey, we may use the concept "understanding" in order to describe an activity in which a given subject approaches an event or an experience such that s(he) or it may experience it by himself in a way which is not mediated by a formal abstract representation, but rather a holistic approach from "within" the phenomenon. When I see a suffering person, I may describe his situation by using a variety of formal abstract models, psychological, sociological or others. In contrast, I may also try to "understand" his situation if there is a process through which I share, to a certain extent, his suffering or recognize it as an experience I am familiar with. Such a process may take place through literary works, socialization, etc., by approaching the suffering person from "within" his suffering. The term "compassion" exactly illustrates a situation in which one shares ("co," the prefix in Latin for joint activity) the passions of the suffering subject from within.

It should be noted that in both cases, explanation and understanding, there is a social aspect. However, while the social in "explanation" is a source of validity that comes after a formal representation was created by a given subject/community, in "understanding" the shared experience is the basic platform and the existential condition for the activity of understanding. In this book, I clearly differentiate between explanation and understanding and follows Dilthey's method in my own inquiry.

Concerning evolutionary theory and language, the case is quite clear. In contrast with powerful scientific explanations such as Newton's laws, the evolutionary explanation of language is far from being an agreed scientific explanation for human being's unique symbolic activity and to model the way it has evolved out of simpler forms of communication. It is neither a method that leads us to "understand" a given experience similar to the process of literary works that cause their readers to identify with their heroes or to the process that causes mathematicians to be excited by the beauty of a mathematical proof. Therefore, and in the current state of affairs, we have good reason to conclude that we cannot better explain or understand the phenomenon of symbols by the Darwinian evolutionary mechanisms per se, and that some kind of a *social* narration must be taken into account, if not as an explanatory device, then as spectacles for better understanding certain portions of our semiotic experience. The next chapter aims to approach semiotics from a social perspective. This chapter is only the first step in understanding signs through a social-somatic-semiotic perspective that aims to locate our semiotic ability in the context of both body and society. However, before approaching the next chapter, let us read an excerpt from Barth's novel, "The End of the Road." This excerpt mocks any form of brute and over-inclusive reductionism and clearly applies to the linguistic pretensions of the Darwinian evolutionary theory:

> The dance of sex: If one had no other reason for choosing to subscribe to Freud, what could be more charming than to believe that the whole vaudeville of the world, the entire dizzy circus of history, is but a fancy mating dance? That dictators burn Jews and businessmen vote Republican, that helmsmen steer ships and ladies play bridge, that girls study grammar and boys engineering all at behest of the Absolute Genital? ... Who would

not delight in telling some extragalactic tourist, "On our planet sir, males and females copulate. Moreover, they enjoy copulating. But for various reasons they cannot do this whenever, wherever, and with whomever they choose. Hence all this running around that you observe. Hence the world?" A therapeutic notion. (Barth, 1988, p. 341)

The lesson in this excerpt is self-evident. Understanding can never be achieved through simple reductionist models. When we deal with the human, who is clearly complex, we should understand him through weaving the threads of the somatic, the semiotic and the social that creates this unique fabric.

Cat-logue 2
Are There Jewish Cats?

Bamba: I was quite disappointed to read your opinions concerning the evolution of language.
Dr. N: Do you mean "opinions" in the sense of those artificial things measured in public opinion surveys? Well, in this case I don't have opinions. I never hold opinions or attitudes in the static unidimensional sense used by social psychologists. In any case, I am sorry to disappoint you. But why are you so disappointed?
Bamba: First, because you sound like a creationist and I don't like those anti-scientific guys. Second, because I had hoped that we were family, as the Darwinian theory suggests.
Dr. N: Do you mean in Wittgenstein's sense of *family resemblance* or in the biological sense of *biological resemblance*? In the first sense, we are relatives. Do not forget that Jews have been portrayed in Medieval Europe as demonic creatures with tails. So if you have a tail and I have a tail, then there is some family resemblance!
Bamba: I did not know that to have a tail in Medieval Europe was such a shame! It seems to me to be a great honor.
Dr. N: A shame for some people of course. You should remember that *animalism* was a popular rhetorical tactic at those times. Instead of arguing with someone or confronting his otherness, you might describe him as an animal and in this way to dehumanize him and delegitimize his existence.
Bamba: Interesting. Are there Jewish cats?
Dr. N: Hmmm...it is difficult to answer that question because it involves a *categorical error*. "Jewish" is a category that concerns human beings and not animals in the same way as "biology" involves biological entities and "physics" involves physical entities. You see, this categorical error is what so upset me concerning the Darwinian evolutionary theory of language.
Bamba: But let us return to our discussion. Do you think that we don't resemble one another biologically?
Dr. N: Well...there is convincing scientific evidence that you and I are quite similar. I mean, not as similar as a chimpanzee and I, but similar to a lesser extent. My argument is not against the Darwinian evolutionary theory in general, but specifically against the evolutionary theory of language. Evolutionary theory is just an explanatory device and not a portrayal of reality. It is a model and a model is about something. It is not the something in itself. Please keep this in mind because it seems that you have fallen into this pit like many other decent people...I mean cats of course. Since a model is just a model, we should always examine how successful it is in explaining to us a given portion of the world. In our case, we may accept the evolutionary theory in one place, such as in the

biological realm, and reject it in another place, such as in the semiotic realm.
Bamba: Hey, but since there are so many indications that support evolutionary theory in general, why don't you want to accept it for the particular phenomenon of language?
Dr. N: You see, Darwin's theory of evolution is based on fossils and we do not have fossils of our minds! We may assume that we are relatives of animals, and then compare our mental faculties with the mental faculties of other animals. However, this way of inquiry is logically invalid because we first assume that we are relatives, then compare our faculties, and then what? Prove that we are relatives?! This is a kind of vicious circularity any serious man (or cat) should avoid.
Bamba: Do you have evidence that the evolutionary explanation of language is false?
Dr. N: No.
Bamba: Therefore, you must conclude that it is correct!
Dr. N: Remember that just because a theory has not been proven false, we cannot conclude that it is true. This line of reasoning is clearly false and it is described in logic as *Ad Ignorantiam*, appeal to ignorance.
Bamba: So, what is your suggestion? Has language evolved out of something? Have people always had the ability to communicate through abstract signs?
Dr. N: Well, I don't have certain answers to your questions, but I believe that social is a key term in trying to answer these question. In this context there is a very interesting idea that our sign systems have evolved *culturally* from concrete material experience in the world into a very abstract system of signs. I do not accept this idea, but this narrative is worth discussion.
Bamba: Do you use the term "narrative" in order to increase your sales among Post-Modernist readers.
Dr. N: This time you got me.
Bamba: Do you think that cats are included in this "narrative" of yours?
Dr. N: Unfortunately not. Nevertheless, remember that the idea of cultural evolution is a Marxist idea.
Bamba: That draws on the philosophy of Hegel.
Dr. N: Correct, and Marx, although a Jew, was not a cat.
Bamba: Neither was Hegel.
Dr. N: A Jew or a cat?
Bamba: A Jew of course! Concerning cats, I'm not sure. Some of Hegel's ideas are so obscure that they seem as if they had been written by a rutting cat.
Dr. N: Well, don't be so harsh with your rutting peers. Remember that our minds are not only immersed in the social but in the body and its practice. Even human beings find it difficult to develop highly abstract and clear philosophical theses when their minds are distorted by acute bodily needs.
Bamba: So what is next?

Dr. N: Since we mentioned the term "social" as a key for understanding our sign system, let us elaborate on this thesis by presenting Saussure's and Volosinov's ideas of language as a social system.

Chapter 6
Saussure and Semiotics as a Social System

Summary: In contrast to European philological research that tries to trace the origins of language, Saussure suggests that we should study language as an abstract system of signs, and that language is primarily a social system.

The idea that language should be studied as a social system is closely associated with the name of Ferdinand de Saussure. The linguistic research Saussure was familiar with traced the phases of linguistic form evolved to reach its present state. This philological-historical research, I previously noted, was described by Saussure as a *diachronic* approach to linguistic research. Saussure defines diachronic linguistics as the study of "relations that bind together successive terms not perceived by the collective mind but substituted for each other without forming a system." He also argues that for the linguist who describes language from the "inside," as it exists for its users, the diachronic approach is of no use, since history means nothing to the users of language in the present. Whether a specific word has evolved out of this form or another may be of interest to the outside observer, but it has no relevance to those who use words for communication:

> The first thing that strikes us when we study the facts of language is that their succession in time does not exist insofar as the speaker is concerned. He is confronted with a state. (1959, p. 81)

His conclusion is that the linguist

> can enter the mind of the speakers only by completely suppressing the past. (p. 81)

Saussure explains this argument by analogy to the game of chess. Each chess piece has its semiotic value, which depends on its position on the chessboard. The game is governed by laws, which are the "constant principles of semiology." In order to understand the current state of the game one does not need to recall the game's history (e.g., that the game originated in Iran), but to be familiar with the rules of the game and to analyze the current arrangement of the pieces on the board. The arrangement of signs in language is what constitutes the rules of a *synchronic* approach to linguistics. This, of course, was a revolutionary linguistic idea since it suggested examining language by ignoring its outer layer and focusing on the abstract relations that constitute the *system* under inquiry. This ideal of synchronic linguistics was partially fulfilled by Chomsky's theory of syntax. However, the applications of Saussure's synchronic vision are not limited to syntax, and we may query its relevance for understanding the meaning of sign systems in general.

In order to study language as a "game," one should adopt a synchronic approach that deals with the "logical and psychological relations that bind together coexisting terms and form a system in the collective mind of speakers." By that Saussure means that we should consider language as a *network of relations* in which the value of given signs (= words) depends on the value of the other signs in the net.

At this point, it is important to dwell a little bit more upon Saussure's unique conception of the sign because it may shed some light on the issue of abstractness previously discussed. According to Saussure, a sign is a synergetic product of a concept and a sound pattern, an idea (a concept) and its concrete manifestation in speech (or in another medium). This unique conception of the sign contrasts with the idea that signs function as names for objects or properties already in advance of language (i.e., the nomenclature conception). This idea brings us back to the Garden of Eden by reminding us that the nomenclature conception of signs already appears in the Book of Genesis:

> And out of the ground the Lord God formed every beast of the field, and every fowl of the air; and brought them unto Adam to see what he would call them: and whatsoever Adam called every living creature, that was the name thereof.

According to the semiotic thesis presented in the Book of Genesis, things have a true essence and names simply reflect this essence. This point of view, which has dominated linguistic theories for years, considers words to be very closely associated, and even closely associated in a mystical sense, with their objects. This conception is also evident in one of Plato's dialogues (i.e., Cratylus), and the convergence of the biblical source with the later Greek source may lead us to the conclusion that either the divine source of knowledge has been spread in different cultures or that there is a kind of a naive theory of signs (a Jungian semiotic archetype?) that results from the universal experience of human beings. Whether there is a universal and naive theory of signs is an interesting question. The evidence concerning naive theories across cultural settings may support this possibility. It must be noted, however, that at least in Plato's dialogues there is the competing conception to the nomenclature conception! A semi-Saussurian conception, suggesting that names are just arbitrarily associated with their objects. Therefore, if we assume some kind of a naive theory of signs, it should include the rival naive theory too! Well ... things are always more complicated than we usually think and whenever we jump up and declare that we have uncovered a hidden structure of reality, another contradicting structure pops up out of the blue, as if reality is enjoying refuting our naive structuralism and teaching us once again an important lesson concerning the dialogical and argumentative nature of our mind. Anyway, it seems that both the nomenclature and the counter-nomenclature perspective share the idea that the sign is something that stands for "something else." Is this "something else" an object that precedes its signification? Not exactly. Saussure and other thinkers, such as the later Wittgenstein, have a different idea.

In order to present this idea, let us recall that for Saussure the sign is a mental association between a concept and its material expression. The question is, what does the concept stands for. Is it a concept that corresponds to a thing in the reality outside the subject or inside the mind of the subject? Saussure's answer dismisses the referential power of signs and suggests that the sign neither corresponds with an external object outside the subject nor signifies an internal idea/concept in the mind of the subject, but derives its *meaning* from the *system of signs* of which it is a part.

This counter-nomenclature concept of meaning has a far-reaching implication for the relation between language and thought since it suggests that thought, rather than reflecting inner/outer reality, is a *semiotic phenomenon and that without a linguistic structure, thought as we know it simply does not exist*:

> Psychologically our thought - apart from its expression in words - is only a shapeless and indistinct mass. (1959, p. 111)

and

> There are no preexisting ideas, and nothing is distinct before the appearance of language. (1959, p. 112)

We may examine this far-reaching statement in the context of non-human organisms: Does Saussure mean that chimps do not think? Moreover, if they express cognitive abilities we may describe as "thinking," in what sense are those abilities different from the concept of thought as used by Saussure?

Luria and Vygotsky (1992), who contest Saussure's identification of thinking with language, suggest that thinking and speech have different roots and that at the early phases of development one can exist without the other.[23] Indeed, if thinking is defined as an intellectual attempt to solve a problem, then apes (and other organisms) manifest thinking without speech as do young children. On the other hand, primitive forms of signification, such as shouts, laughing or crying can exist without any relation to thinking. This separation between language (speech) and thought is not limited, of course, to children or apes and it is evident in the psychological experience of adults. In this context, Luria and Vygotsky say something very interesting about language (or sign systems in general). They suggest that the child discovers the functional use of words as a means of communicating with others, and then at a certain developmental phase turns inward, internalizes the use of language, and uses language as a tool with which to think. For example, imagine a three-year-old child who tries to put a glass on the table. His father may instruct him "go slowly, hold the glass with both of your hands, very nice, now raise your hands and put it gently on the table." This regulatory procedure conducted by the father is internalized and the child acquires through internal speech a sort of regulatory mechanism that primarily exists at the social level. As already suggested by von Humboldt (1992): "Apart from the communication between one human and another, speech is a necessary condition for reflection in solitude" (p. 101). This statement prompts us again to examine the validity of our reified universe, because it seems to emerge out of our semiotic reflection and not from our pre-semiotic encounter with reality.

Luria and Vygotsky suggest that certain forms of thinking may exist without language (like mental imagery), but that language (or semiotic systems in general) is a unique and powerful tool with which to think.[24] Following Peirce, we realize that there is no thinking (cognition) without mediating signs. Therefore, Luria and Vygotsky's argument should be qualified and rephrased in terms of different levels and forms of semiotic activity that differentiate between different cognitive activities. In addition, the idea of internalizing language from the outside (the social) to the inside (the psychological) seems to create enormous problems we may easily avoid. In the next chapter I presents Volosinov's semiotic theory that clearly undertakes the challenge to transcend the dichotomy between the subject and the social.

Let us return to Saussure. The answer Saussure (but also Volosinov and later Wittgenstein) provides both to the nomenclature challenge, and to the difficulties concerning sign-signified relations, is to consider the sign through a *social-systemic perspective of communication*. For example, in his famous building example Wittgenstein (1968, p.3) describes the linguistic exchange of signs between two builders:

> Language is meant to serve for communication between a builder A and an assistant B. A is building with building stones: There are blocks, pillars, slabs and beams. B has to pass the stones in the order in which A needs them. For this purpose they use language consisting of the words "block," "pillar," "slab," "beam." A calls them out; B brings the stone which he has learnt to bring at such-and-such a call.

The above example raises the question whether the meaning of the signs A and B use is determined by the correlation between the linguistic sign and its concrete reference in the outside world or by the communication process that determines the meaning of the signs.

According to Saussure and Wittgenstein, *the social communication process is the one that determines the correlation rather than the correlation determining the communication process*. In this sense, the meaning of a sign cannot be determined by looking outside of the linguistic system for a well-defined reference in the outside world

Saussure and Semiotics as a Social System

(since, as we have already realized, such a reference does not exist), nor inside the mind, but rather by getting into the system and examining the unique position a specific sign has within a net of signs.

At this phase we can see that Saussure's definition of language ("the whole set of linguistic habits which allows the individual to understand and to be understood," p. 77) points at the social-communicational meaning of language. The function of language, according to Saussure, is not to mirror our inner psychological states nor to mirror the external world. Language is for communication of the collective:

> If we could embrace the sum of word-images stored in the minds of all individuals, we could identify the social bond that constitutes language. It is a storehouse filled by the members of a given community through their active use of speaking, a grammatical system that has a potential existence in each brain, or, more specifically, in the brains of a group of individuals. For language is not complete in any speaker; it exists perfectly only within a collectivity. (1959, pp. 13-14)

Another important component of Saussure's semiotic theory concerns the *arbitrariness* of signs. By arbitrariness he means that for any given language the choice of actual signs from among the range of possible signs is entirely unconstrained by internal constraints of the linguistic system or constraints external to the social institution of *La Langue* (Harris, 1987). According to this idea, a sign, as a material entity, is only arbitrarily associated with its meaning, the same as the outer layer in which the king in a chess game is associated with its function. Since the material expression of a sign is only arbitrarily associated with the concept, we cannot disclose the meaning of a sign by examining its concrete manifestation. This conclusion leads us again to the social aspect of language: Language as a synchronic sign system is meaningful only as a social fact and should be studied as a social phenomenon. This argument already appears at the beginning of "Course in General Linguistics," when Saussure states that "language is a social fact" (p. 6).

The arbitrariness of the signs leads us to one of the interesting aspects of Saussure's systemic conception of signs: the way he conceptualizes the *identity* of a sign. Saussure discusses the question of identity in the context of synchronic linguistics. He raises the question of what causes us to name a street that has been completely rebuilt by the same name we used before the street was rebuilt. After all, the new street may be totally different in terms of its material properties from the old street. His answer is that the street, which is the subject of the identity relation, is not a material entity, but a position within a system of streets. *In this sense, the identity of a sign is determined by its position (context?) rather than by some essential properties of the object, which this symbol signifies.* This position explains the way a sign may acquire a certain value, a certain sense of identity (i.e., a cat is a cat) without the need of pointing at an object preceding its signification (i.e., the "real cat").

The arguments presented so far sharply demarcate language as an abstract and fixed, logical sign system (*La Langue*) from the physical expressions of *La Langue* is apparent in speech (*La Parole*). By differentiating synchronic from diachronic linguistics, Saussure was trying to explain language as a closed system with its own terms without using exogenous factors, such as history, as characterized the dominant linguistic research that preceded him.

The question of what aspects of human language should be considered "synchronic" and what aspects of language should be considered "diachronic" is a matter of dispute. Saussure acknowledges the difficulty in differentiating between the two systems in practice:

> Speech always implies both an established system and an evolution; at every moment, it is an established institution and a product of the past. To distinguish between the system and its history, between what it is and what it was, seems very simple at first glance; actually the two things are so closely related that we can scarcely keep them apart. (1959, p. 8)

The difficulty of differentiating between the synchronic and diachronic facets of language is not the only problem in Saussure's theory. However, this is a problem of which Saussure was well aware and which he explicitly mentions in the text. Other problems do not receive the same treatment. For example, the social meaning of language and its relation to language as a static system is far from clear to the reader. Since Saussure's seminal book ("Course in General Linguistics") was edited by his disciples out of his lectures, it is difficult to consider it a coherent and systematic text that gives satisfactory explanations to its major ideas. Saussure's ideas had an enormous influence on semiotic research and several semioticians took up the challenge of elaborating the idea of a *social theory of semiotics*. Among those semioticians was Valentin Volosinov. The next chapter is devoted to his socio-semiotic theory of the mind. But before approaching Volosinov's socio-semiotic theory, let me summarize the current chapter. After presenting the historical theories of language and their shortcomings in approaching the phenomenon of the symbol, we turned to examine language from a social perspective. We presented Saussure's conception of the sign system as a closed social system and examined the implications of this idea for meaning, identity, and the relationship between mind and reality. We found that Saussure explained the way in which signs may acquire meaning and a sense of value (i.e., identity) without pointing at a real object. Again, we failed to explain the way the abstract/arbitrary symbol is related to our somatic experience and the way our semiotic ability leads to the reification of the world. However, we learned that the *social* and the *systemic* aspects of semiotics may pave the way for better understanding of this mystery. The next chapter attempts to establish another layer in this process of understanding.

Chapter 7
The Mind as a Semiotic Interface

Summary: Volosinov presents a radical theory of the mind: The mind is a semiotic phenomenon that rests on social grounds. This theory and its relevance to the meaning of the sign is examined in the current chapter.

Volosinov's socio-semiotic theory is presented in his book "Marxism and the Philosophy of Language," written in 1929 but first translated into English in the early 1970s. Unfortunately, this book, one of the greatest intellectual and cross-disciplinary achievements of the twentieth century, went largely unheeded by its relevant scientific audience from the humanities and the social sciences in general[25] and the systems science community in particular. This disregard may be attributed to psychological avoidance by certain intellectual circles of any scholarly work that refers to Marxism and, on the other hand, of Marxist intellectual circles that found no reference to Marx in this book. Indeed, "Marxism and the Philosophy of Language" does not refer to Marx's work, and it is, as Volosinov himself suggested, a pioneering venture, with no substantial roots in Marxist theory. Another possible explanation for the disregard of Volosinov's work may be its anti-reductionist spirit, which sharply criticizes most of the scientific paradigms that still govern our academic life. In this sense, Volosinov's work is of enormous value to the systems science community that adopted an anti-reductionist approach as one of its mainstays.

We know almost nothing about Volosinov the man, and the scientific literature that refers to his work mostly discusses the question whether Volosinov's work can be attributed to the well-known Russian scholar Bakhtin. In contrast to the disregard surrounding Volosinov the man and the problem of authorship, the intellectual context in which he operated and to which he responded is well known.

During the 1920s, the semiotic theory of Saussure, previously presented, was a major source of interest for Russian intellectuals and the separation between "*La Langue*" and "*La Parole*" was a major controversy among those intellectuals circles. Volosinov aimed to bridge the specific gap between the two apparently different systems, but more important to create a system of thought that *transcends other dualistic relations* as well. As suggested by Matejka:

> In his attempt to operate as a dialectician, he sees individualistic subjectivism and abstract objectivism as thesis and antithesis and proposed as dialectical synthesis beyond this opposing trends, a synthesis that would

constitute a negation of both thesis and antithesis alike. (in Volosinov, 1986, p. 169).

Matejka's interpretation of Volosinov's work locates it in a Hegelian-Marxist tradition. My own reading of Volosinov is quite different and locates his thesis closer to radical scientific notions of complexity (Neuman, in press). Complexity is an issue I aim to discuss later, but the general idea of complexity as a non-linear, multilevel, dynamic approach to the study of multi-agent systems is the one that should guide the reader during this chapter.

The most important theoretical contribution of Volosinov is his attempt to lay the foundations for a socio-semiotic theory of the human mind. The mind, or "psychic experience" as he calls it, is considered as "the semiotic expression of the contact between the organism and the outside environment" (p. 26). In modern terms, we may rephrase this statement by saying that psychic experience can be considered as the *interface* between the organism and its environment. This suggestion is important since it portrays *the mind as a boundary phenomenon* and not as a fixed entity that can be reduced either "downward" to a material brain or "upward" to vague metaphysical concepts, such as the Cartesian self. This idea is important since, if we are part of reality, then our sign system (mind) demarcates the boundary between ourselves and a wider system of which we are a part. In this sense, *mind as a semiotic interface must have a recursive-hierarchical structure that corresponds to the fact that it signifies a system of which it is a part and upon which it reflects*.

The idea of a semiotic system as a boundary phenomenon is also important for several other reasons, such as that it presents a radical alternative to the two main narratives of the West: empiricism and rationalism. By portraying the mind as a boundary phenomenon, as the interface between the organism and its environment, one cannot understand the mind by turning outside to the environment, neither by turning inside, but as a unique and differentiated system with its own "language"[26]:

> Outside the material of signs there is no psyche; there are physiological processes, processes in the nervous system but no subjective psyche as a special existential quality fundamentally distinct from both the physiological processes occurring within the organism and the reality encompassing the organism from the outside, to which the organism reacts and which one way or another it reflects. (p. 26)

This excerpt should be carefully read repeatedly because Volosinov is making an important statement. He is not denying the possibility that cognitive processes can take place apart from semiotic activity but that "no subjective psyche as a special existential quality" exists outside the realm of signs. This statement is in line with the position presented by Luria: Thinking can take place without semiotic mediation, but semiotic mediation gives the human mind its uniqueness. Again, this suggestion should be qualified since, as we learned from Peirce, there is no cognition without signs.

Now, the next question we should deal with concerns the origin of the sign systems. The origin of signs can be located in the mind, as some rationalists have tried to argue. According to this explanation, we have some innate capacity to think in signs. We are "hardwired," and we use a brain characterized by symbolic activity. Volosinov considers this explanation to be a fundamental epistemological error because signs are those that constitute the mind and not vice versa:

> Consciousness [= mind] becomes consciousness only once it has been filled with ideological [semiotic] content... (p. 11)

and

> In other words, the individual consciousness not only cannot be used to explain anything, but, on the contrary, is itself in need of explanation from the vantage point of the social, ideological [= semiotic] medium. (p. 12)

Thus, the mind is a concept that cannot explain anything, a point also raised by Vygotsky, but which needs an explanation in itself. The explanation provided by Volosinov is that the mind is the semiotic activity that cannot be reduced to the brain or to the external environment.

After pointing at signs as constituting the mind and considering the mind as something that needs an explanation rather than as a mysterious explanatory device, Volosinov turns to discuss the origins of the mind as a semiotic system. He argues that both materialists' and idealists' conceptions of the mind do not provide us with sufficient answers to the question of what is mind and what are the origins of the mind as a sign system. His suggestion is that "Signs can arise only on *interindividual territory"* (p. 12) and that the mind as a semiotic activity is grounded in a social activity of communication. By adopting this position, the mind is portrayed as a multilevel semiotic phenomenon that stretches beyond the boundaries of the individual skull to the social realm of communicating agents: "From the standpoint of content, there is no basic division between the psyche and ideology [i.e., a semiotic system]; the difference is one of degree only" (p. 33).

This conception of the social aspect of the semiotic system causes Volosinov to oppose the Saussurian distinction between synchronic and diachronic forms of linguistics. Volosinov considers Saussure's views as a high point of abstract objectivism and rationalism. According to Saussure, language is social in the sense that it is a stable system exterior to the individual consciousness. In order to rebut this stance, Volosinov presents the following question: To what degree may the system of language be considered a *real* entity? Since this system does not have a concrete material reality (such as physical systems), we must conclude that it exists only in the minds of individuals belonging to some particular community governed by norms. Since norms change, unless we are speaking of "linguistic norms" imprinted in the minds of the individuals, what is the meaning of "objectivity" or "real" concerning the linguistic system? Volosinov's suggestion is that the objectivity lies at the *relationship* between the individual minds and the linguistic norms. That is, the norms in themselves are not objective facts but the relationship between the collective and the individual mind is. In this sense, Volosinov's ideas remind us of Vygotsky's law of genetic development stating that the individual consciousness is a reflection of the social sphere. Volosinov points at the fact that this important idea has slipped the mind of the rationalists who have severe difficulties in trying to figure out the reality that language as an "objective" system does possess. In this sense:

> Language as a system of normatively identical forms is an abstraction justifiable in theory and practice only from the standpoint of deciphering and teaching a dead alien language. This system cannot serve as a basis for understanding and explaining linguistic facts as they really exist and come into being...this system leads us away from the living dynamic reality of language and its social functions. (p. 82)

After presenting Volosinov's general semiotic ideas of the mind as a socio-semiotic system, we may turn again to the sign and to the way the meaning of the sign is determined. According to Saussure, language is a closed system of meaning, and the question is how language is related to the world. Similarly to Saussure, Volosinov suggests that the understanding of the sign is not determined by its relation to a real object out there in the external world, but through a network of other signs: "The understanding of signs is after all an act of reference between the sign apprehended and other, already known signs" (p. 11). In this sense, understanding a sign is possible only within a given system of signs and as a part of a social process, since "Signs emerge, after all, only in the process of interaction between one individual consciousness and another" (p. 11). Therefore:

> what is important for the speaker about a linguistic form is not that it is a stable and always self-equivalent signal, but that it is an always changeable and adaptable sign. (p. 68)

Since, in other words, the social activity of communication is the only framework for understanding the use of signs, and not the abstract and static system of language, *the signs in our language are not stable and self-identical entities corresponding to some "real" objects,* but changeable and dynamic things. This is an important argument since all the objects in our world are evident through signification. However, if our signs are dynamic things that correspond to some social dynamic rather

The Mind as a Semiotic Interface   51

than to external reality, how is it possible to approach the world in static terms? Let us keep this important question in mind while proceeding.

Volosinov contrasts signs (symbols) with signals that have a relatively fixed function and exist among animals; whenever a signal is produced it always refers to the same object or serves the same function. Volosinov further explains this point by differentiating between two activities: *recognition* and *understanding*. While animals recognize things, human beings *understand* them, and: "Only a sign can be understood; what is recognized is a symbol" (p. 68).

What follows from this analysis is that the sign system we use is cognitively economic in the sense that the same limited number of signs may serve to communicate different things across different contexts. When I say "pig" I may refer once to a given animal that I see, and on another occasion use the same sign in order to insult a person. Therefore, the question whether the sign points at something outside or just receives its meaning from its position in an abstract system of signs is answered by Volosinov by neither-nor. The unique answer supplied by Volosinov is that the sign is a dynamic process that does not have a fixed referential meaning (either outside or inside a system), or a static location in a wider system of signs, and that *understanding*, which is the emerging dynamic out of semiotic activity and not recognition, takes place only in a particular context of communication between interlocutors:

> Thus the constituent factor for the linguistic form, as for the sign, is not at all its self-identity as signal but its specific variability; and the constituent factor for understanding the linguistic form is not recognition of the "same thing" but...orientation in the particular, given context and in the particular, given situation...orientation in the dynamic process of becoming.... (p. 69)

So what are understanding and meaning according to Volosinov?

> Any act of understanding is a response, i.e., it translates what is being understood into a new context from which a response is made. (p. 69)

Moreover, what is the most basic unit of understanding? Is it the sign? At this point Volosinov provides a surprising answer: *In itself the sign does not have any meaning.* In order to understand this radical conception, let me try to reconstruct Volosinov's argument. Volosinov suggests that, by definition, meaning is a property of the *utterance* (the most basic unit of communication) as a whole. He describes this meaning as the theme of the utterance. The *theme* is an individual, concrete, context-dependent and irreproducible act of communication. For example, consider the question: "What time is it?" This question has a different meaning (in the sense previously described) depending on the specific *context* in which it is uttered. From this analysis it follows that the meaning of the utterance, the theme, is dependent not only on the linguistic forms of which it is comprised it (words, syntax, etc.), but on extra linguistic factors we may describe as *"context."* The theme is not exactly identified with the meaning of the utterance and it may be better described as "a complex, dynamic system of signs that attempts to be adequate to a given instant of generative process." This sense should be differentiated from meaning which is the "technical apparatus for the implementation of a theme" (p. 100). In this sense, the fact that the words composing the question "What time is it?" have common-normative "meaning" allows us to generate the theme with all its particularity in a given context. Therefore, the conventional meaning of signs can be served only as a *technical and consensual (social) platform* for the generation of meaning (theme = response) in a concrete context. Volosinov stresses the fact that we cannot draw an absolute boundary between theme and meaning of an utterance and that there is no theme without meaning and vice versa. However, he stresses the idea that only theme means something and that meaning (in the abstract Saussurian sense of the term) "means nothing." The sign means nothing, but only as a conventional platform for a theme, unless it is understood as a part of a *dynamic process of interaction within a given context*.

Volosinov's semiotic theory supplies us with some answers to questions raised in the previous chapters, and with some insights concerning the way our inquiry may proceed. If we may summarize Volosinov's theory in a couple of words, we may say that this is a theory that suggests that the mind is a unique *semiotic phenomenon* that exists as a *boundary* between the organism and its environment. It is a *process of signification* that can be considered as *emerging* from *social interaction* and can be explained only as a social phenomenon. In this context, and except for abstract scholarly analysis, *signs should not be considered to be fixed entities* that obey the law of identity (A=A), or signify inside/outside objects. *Signs are conventional, indeterminate and dynamic platforms for the emergence of meaning (= response, understanding) in a local and social context of interaction.*

Moreover, signs receive their significance from the whole utterance in which they are components (from the macro-level of analysis) and the whole utterance receives its significance in a circular and spiral-like process both from the components that compose it (micro-level) and from the wider context in which it is included. The idea of a spiral-like process of understanding is the motif of this book. Keep it in mind while we proceed to the next chapters. Also keep in mind that Volosinov fails, like Saussure, the European classical philologists and the Darwinian linguists did, to deal with the symbol-grounding problem since his theory ignores the fleshy basis of our existence. In this sense, he avoids dealing with the development of the sign system and with its embodied nature.

Cat-logue 3
It Means Nothing

Bamba: Good morning to you...did I tell you that there is a huge, disgusting dog sniffing around our yard?
Dr. N: What do you mean by a "dog"?
Bamba: Aha! Trying to get even and to force me to admit the existence of an object! You know exactly what I mean by "dog" because you just wrote about language as a social system. According to Saussure the meaning of the sign "dog" is determined by its unique position in our semiotic net. A "dog" is not a "cat," it is not a "cow" and it is not a "shoe"!
Dr. N: You see that the problem is that this list can be infinitely long, because a "dog" is neither a nose nor a goat, and neither is it infinitely many other things. Therefore, the Saussurian method for determining the meaning of the sign is quite problematic. Do you think that the abstract sign "dog" you used in your utterance is sufficient in order to determine the whole meaning of your utterance? I mean... according to the Saussurian view the utterance, "I love mice" should have the same meaning whether it is produced by a fellow mouse or by a hungry cat.
Bamba: Well, I think that you do not understand Saussure's argument. However, concerning the specific example you gave, personally I can testify that it has a very different meaning.
Dr. N: So, following Volosinov, if we consider language to be a communication system, it seems that the *concrete* takes precedence over the *abstract*, the *whole* over the *elements* and the *dynamic* over the *reified*. Keep it in mind because those ideas are central to the holistic and dynamic thesis I present in this book.
Bamba: But in both cases, a mouse is a mouse. Isn't it?
Dr. N: Following Volosinov I tend to believe that the meaning of a sign is determined entirely by the whole utterance which is embedded in a significant context. Saussure's idea that we are walking around with some shared dictionary in our heads seems quite bizarre to me.
Bamba: I suspected a long time ago that you are some kind of a Post-Modernist thinker! Do you suggest that the sign "dog" has no meaning?
Dr. N: I assume that you refer to some kind of a *structural stability*. In this sense, the sign "dog" has material stability. Within the same language it sounds quite similar across different contexts and usually refers to members of what we may describe as the dog class. However, Volosinov is saying something different, and very interesting, about meaning. He is saying that the meaning of a sign means nothing, it only possesses potentiality – the possibility of having a meaning within a concrete theme.
Bamba: Hmmm...interesting and it sounds quite clever for a human being. What does he mean by "theme"?

Dr. N: Volosinov uses the term "theme" to describe the significance of the whole utterance and defines it as "a complex dynamic system of signs that attempts to be adequate to a given instance of generative process." You see now that understanding an utterance is an active process and neither a passive recognition of some elements in our conceptual dictionary nor in our external environment. This is why Volosinov suggested that "meaning belongs to a word in its position *between* speakers" and that it is "the *effect* of interaction between speaker and listener produced via the material of a particular sound complex."

Bamba: After this short exposition on Volosinov's semiotic theory of meaning, try explaining to me the meaning of the utterance that opened our discussion: "There is a huge, disgusting dog sniffing around in our yard."

Dr. N: If I correctly analyze your evolutionary position with regards to dogs, and your personal painful history with this kind, I believe that what you are trying to say is that you are scared to death of the dog and that you want me to throw it out of the yard.

Bamba: I believe that I can handle the dog by myself, but since I don't want to dispute your scholarly interpretation, I tend to accept it...Please get rid of this dog before he bites my tail!

Chapter 8
We Have Never Been Too Abstract

It is true, as Marx says, that history does not walk on its head, but it is also true that it does not think with its feet.
Merleau-Ponty

> Summary: If the dynamic, the concrete and the whole stand at the heart of our sign system, our mind and our unique form of being-in-the-world, how is it that our universe is conceived as reified and fragmented? A possible answer is that we have undergone a form of cultural evolution that transformed us from a concrete, holistic and operational form of existence to an abstract form of existence. This idea of cultural evolution is examined in the current chapter.

Whether we evolved from simpler forms of being or were created by God, whether we have evolved in a continuous line of development, through a discontinuous genetic mutation, or through a Catastrophic Event that combines the continuous-discontinuous idea, the dawn of human cognition is a mystery and, excluding wild scholarly speculations, our ability to clarify this mystery is severely limited. However, the epigenesis of the human mind has been described from both psychological, developmental and cultural perspectives as striving towards the abstract. There are very good reasons to conclude that at least on the collective cultural (macro) level of analysis (and not only the individualistic one) the human mind has gradually detached from its concrete, dynamic and holistic experience with the world and moved toward an abstract, static and fragmented form of existence. If this is the process, then the reification of the world can be explained by this development from a pre-objective, somatic, dynamic and holistic encounter with reality to a realm of abstract and fragmented signs. As will be immediately illustrated, this striving for the abstract seems to appear in apparently diverse semiotic phenomena such as "knowledge" and "money."

The most ancient metaphor of knowledge consumption is the physical organic act of incorporation through eating or sexual intercourse (Kilgour, 1990). For example, in the Book of Genesis the verb "to know" appears in the context of eating and sexual intercourse. Those activities are very similar in the sense that both involve the physical activity of moving something from the outside to the inside. Adam and Eve, who ate from the Tree of Knowledge, and Jesus, who shared his spirit with his disciples through the symbolic act of the communion, are just some of the most famous examples of knowledge consumption through physical (or organic) ingestion. This material basis of knowledge not only characterizes Western civilization. Just think about the African hunters of the Maasai tribe who eat the lion's heart in order to become courageous. This is knowledge, but not the abstract kind we are familiar with in the modern information society. In this material sense, knowledge exists outside the subject as a concrete, physical substance and is literally incorporated through an organic process of assimilation. On the cultural level of analysis, it seems that the concept of knowledge has been historically transformed from a *somatic activity* into an *abstract entity*, the same as the infant proceeds from a material interaction with her environment to interaction with abstract objects.

The notion of knowledge as an abstract entity reached its peak in the modern era through the emergence of advanced information technologies. As described by Miller (1983), the mind came to the modern era "riding on the back of the machine," and new information technologies, such as the electronic computer, have supported schemes through which knowledge can be framed in a symbolic manner dissociated from its concrete nature.[27] *Knowledge, rather than being part of an intimate somatic process located within the social biological activity of human beings, has become detached from human practice and is now conceived of as a universal symbolic language that is independent of dynamic and concrete context.*

A strikingly similar process of cultural development is evident concerning the phenomenon of money. Money, as a ritual object, always represents or signifies something other than itself (Crump, 1981). In this sense, money may be regarded as a semiotic phenomenon,[28] the same as natural language and knowledge. As a symbolic phenomenon, the similarities between money and symbolic knowledge may strike us as overwhelming. Both money and knowledge have undergone similar processes from which they have emerged united: Each, once embedded in a concrete matter (such as gold, silver or copper in the case of money and food and intercourse in the case of knowledge), has become a purified symbolic notion that exists either in the "mind" of modern stock market computers (in the case of virtual money) or within the "computational" mind of people (in the case of knowledge or information). Even historically most countries have converted to a monetary system that lacks direct correspondence to a concrete reference in the twentieth century (Chown, 1994; Galbraith, 1975) where modern information technologies and the notion of symbolic knowledge established their dominance and celebrated their detachment from reality.

In addition to their similarity on the phenomenological and evolutionary levels, money and symbolic knowledge are limited by the same constraints. Crump (1981) points at two restrictions inherent in money. The same constraints are evident in other semiotic phenomena as well. The first constraint is that money must be capable of circulating indefinitely among those who use it in order to constitute its identity. In contrast with material commodities, the symbolic nature of money holds an idealistic-Platonic phenomenological status, which grants it eternal life as long as it is exchanged. In this sense, virtual money is primarily an *ongoing process* that constitutes through its dynamics a *stable appearance of value and stability*. This point is very important since a strikingly similar process seems to characterize other semiotic systems in which the sign seems to acquire a stable appearance of value while we are well familiar with its dynamic nature and contextual meaning. It seems as if the sign exists, as a stable object, as long as it is on the move. To recall, signs exist only when they are exchanged through a social process of communication. In this sense, they do not have "real" existence outside the appearance of language and exchange. Therefore, the idea that signs-ideas are eternal, static ideas does not hold within our analysis and it seems that it has something to do with the social aspect of the semiotic systems.

The second constraint is that true currency has such a distinctive identity that it has no significant use for non-monetary purposes. In this sense, both money and knowledge, as two examples of signs, have meaning only within a closed semiotic system. That is, the process of semiotic mediation at its most abstract level involves a loose relation between the sign and the signified, and receives its meaning from existing as a communicational phenomenon in the context of human interaction.

The semiotic meaning of money has been discussed by Jean-Joseph Goux in his book, "Symbolic Economies" (1990). Goux suggests that different semiotic phenomena have been transformed according to the same logic:

> A process that is fundamentally isomorphic to the institution of money but of greater complexity may be observed in the genesis of language and the concept – the genesis of the generic term. With the layering of simple operations in several stages and parallel complex procedures combined and the whole branching out in all directions, there results a complex pyramidal organization whose *polymorphous base* is made up of all the world's perceptible and concrete signs and whose summit is composed only of linguistic signs and concepts. That the *genesis of the concept* obeys a logic of substitution, and thus of exchange in the general sense, becomes quite clear in Hegel's analysis: It is a process in which the thought leaves aside what is fortuitous about the thing, the immediate phenomenon, and separates the inessential from the essential, making it into abstraction. (p. 41)

Goux identified three phases of evolution toward the abstract. In the first phase, the sign is embedded in the materiality of the concrete phenomenon. At this phase, the sign is an *icon* of the phenomenon it represents. For example, the earliest form of writing is *pictography*. This form bears an immediate resemblance to the thing or act depicted. It is a phase of simple equivalence. In ethics: an eye for an eye. In commerce: a material good in return for another material good. In semiotics: a word for an object. The second phase, we may describe, following Peirce's "indexicality," involves the extension of the value from its concrete embodiment towards the abstract. It is a process of symbolization, of ideographic writing, of signs that correspond with the phenomena but do not reflect their material-perceptible properties. The last phase is a phase of pure abstraction in which the sign receives its meaning only through its semiotic net. At this phase, the materiality of the sign is only arbitrarily connected to the concept it represents.

Goux, of course, is following Hegel's developmental theory of the mind, although his theory is much more inclined toward the social. Not that Hegel was a Darwinian, but he developed a theory of mental development, which is worth mentioning since it is the basis for Goux's analysis. According to Hegel, man at his most basic form (what we may describe as prehistoric) of existence was a sensual creature who conceived and consumed the world in an unmediated way, in the same way that young children, animals and primitives conceive the world. Man was in the world as an undifferentiated being and not as a distinct subject. Since man in nature is a thinking creature (a creature striving to differentiate and signify), this existential phase was only temporary and man was pushed further to a more distinct form of existence. However, how can man think about the world or himself and at the same time be undifferentiated from it? The answer is that man detached himself gradually from his concrete, sensual experience.

In Inwood's introduction to Hegel's Aesthetic (Hegel, 1993), Inwood illustrates this process through a man who takes a piece of wood and works it into the shape of bison. This act liberates man from his concrete sensual experience in several respects. First, he does not consider the wood as an object to be consumed. Second, he does not chase the wooden bison (although he may worship it), but uses it for communication with others. Third, he does not shape a particular bison, but rather a representative of a bison and, therefore, moves into the realm of the abstract. In this sense, he distances himself from his first relation to the world, the relation of consumption and desire, a primary relation which is also evident in Freud's theory and in Piaget's genetic epistemology, and turns toward being, in our previous terms, a symbolic creature. In terms that have been already mentioned in this book, man has been transformed from an almost undifferentiated existence in the continuous flux of sensation and action into the differentiated existence in the static realm of his sign system.

Hegel develops his thesis in directions we are not forced to accept, such as that the Absolute, that is God, becomes self-conscious in man's cognitive and practical activities. Nevertheless, his idea of evolution is intellectually stimulating, and in contrast to some of Hegel's other wild theoretical speculations even seems to be grounded in the factual realm, as illustrated by Goux.

However, this progressive thesis involves some worrying theoretical difficulties that we may sum up under the title: We have never been too abstract. One of these difficulties is that signs at their most concrete and material form of expression seem to be "abstract" no less than their manifestation by the progressive trio: the adult, the human and the modern. But what does it mean to be an abstract sign? Considering something "in the abstract" means considering it apart from any context of further use, meaning or origin. This interpretation is expressed by Frege who considers abstraction[29] as extraction:

> ...We attend less to a property and it disappears. By making one characteristic after another we get more and more abstract concepts. Inattention is a most efficacious logical concept...Suppose there are a black cat and a white cat sitting side by side before us. We stop attending to their color, and they become colorless, but they are still sitting side by side....We stop attending to their position; they cease to have place, but still remain different. In this way, perhaps we obtain from each of them a general concept of a cat. (quoted in Harrison, 1987, p.55)

This is a nice interpretation of abstraction, and the process of abstraction described by Frege looks like the process of phenomenological reductionism suggested by Husserl. In our case, the question is whether "primitive" people were less able to pay attention to the abstract nature of phenomena than their modern counterparts.

A partial answer to this question may be supplied by Luria and Vygotsky (Luria, 1976), who studied conceptualization among illiterate peasants. Those studies present us with a rich and amusing source of data. Unfortunately, I do not believe that they support Luria and Vygotsky's theoretical insights concerning cultural development, but rather a very different argument concerning rhetorical aspects of language and cultural frames in the sense elaborated by Volosinov and Wittgenstein. Let us see why. One of the subjects Luria and Vygotsky studied was a young illiterate peasant (twenty-two years old) by the name of Illi-Khodzh. Here is the conversation between a researcher and Illi. The researcher's questions are in bold and their interpretation appear in italics:

> **Try to explain to me what a tree is.**
> "Why should I? Everyone knows what a tree is. They don't need me telling them."
> *Rejects need for explanation.*
> **Still, try and explain it.**
> "There are trees here everywhere. You won't find a place that doesn't have trees. So what's the point of my explaining?"
> .
> .
> .
> **How would you define a tree in two words?**
> "In two words? Apple tree, elm, poplar."
> *Enumerates instead of defining.*
> **What is a car? Can you explain it to me?**
> "It uses fire for its power and a person drives it. If it has no oil and no one to drive it, it won't move."

Attempts to define object by citing its features.
How would you explain a car to someone who had never seen one?
"Everyone knows what a car is, there are cars all over the world. There are so many cars it just can't be there are people who have never seen them."
Rejects hypothetical instance.
Say you go to a place where there are no cars. What will you tell people?
"If I go, I'll tell them that buses have four legs, chairs in front for people to sit on, a roof for shade and an engine. But when you get right down to it, I'd say: 'If you get in a car and go for a drive, you'll find out.'"
First tries to define object through graphic description, then insists on the need for firsthand experience. (p. 87)

This amusing dialogue shows us that the peasant is not so stupid, primitive or concrete as portrayed by Vygotsky and Luria. In fact, his answers reflect abstract thinking and some kind of surprise at the questions addressed by the distinguished researchers. For this peasant it seems that those questions are simply *out of context*. If we accept this interpretation, we may conclude that the peasant does not express difficulties in abstract thinking, but difficulties in understanding the *context* in which the question is addressed. This interpretation leads us to reexamine the argument concerning the cultural detachment of the sign from its concrete and material origins. Think, for example, about the pictures of the cave man. Do they look realistic or more like abstract art? Think for example about ancient Egyptian pictures. Do they look like "real" people? Are they less abstract than modern paintings? It can be argued that, like children, the ancient Egyptians were technically less competent in producing realistic pictures and, therefore, their unrealistic pictures result from technical difficulties and not from an "abstract mind." This is, of course, a problematic qualification, because it seems strange that those people who built the pyramids and knew how to mummify people did not know how to paint a real figure. The same is true for the phenomenon of money. Think about the exchange of goods in ancient times. In order to buy a certain good with two cows, three goats and five chickens, the primitive men would have an abstract notion of value no less than his modern counterpart. Therefore, it seems that it is an oversimplification to speak about a cultural development of the sign system and to locate primitives, children and animals at one end of the spectrum and moderns, adults and humans at the other end. However, there is a grain of truth in the cultural developmental perspective. Something interesting really happens when we grow up both as an individual, as a collective, and as individuals in a collective. The change we encounter cannot be disputed, but the way it should be interpreted is quite different. Let us examine a possible perspective that may overcome the critique presented above.

Any semiotic phenomenon may be considered a *structure* or a set of transformations (the exchange of values) that constitute the phenomenon under inquiry. The structure – the transformation/exchange of value – is a stable manifestation embedded in the most basic form of our semiotic ability. This structure is superior to cultural context and time intervals. It is a unique structure of the mind, whether we are speaking about ancient or modern cultures, animals or human beings, children or adults. The ability to use these "abstract" structures of signification characterizes all living systems, from the single cell to the most advanced civilization. However, this abstract structure in itself is limited as an explanatory device. Structures do not have meaning outside a frame and, in our case, it is better to speak about structural transformations that receive meaning from changing *frames* of reference. The ancient monetary system was no less abstract than our monetary system. However, the frame or the context, through which this semiotic system operated and received its meaning, was totally different and therefore created different meaning for the process of exchange, for the meaning of signs, and for the process of thinking in which people were involved. *Therefore, the simple version of cultural development should be rejected in favor of a dynamic theory of micro-macro dynamics.* Adopting this perspective the progressive and linear venture of Hegel and his disciples should be abandoned in favor of a dynamic semiotic theory in which the basic semiotic structure, as subjected to a variety of frames, produces the phenomenological world with all its complexity. This theoretical move is important since it suggests that our reified universe is not only the expression of our symbols and the way they have been transformed from icons to symbol. The next section aims to illustrate further the importance of examining structures and metastructural dynamics for understanding our semiotic system and as another step toward understanding the reification of the world.[30]

8.1. THIS IS NOT A PIPE

Much ink has been spilled on the relation between the sign and the signified and especially about our difficulties in elaborating this relation and considering it through the dynamic of structure and metastructures. I decided to illustrate this difficulty through a painting by Rene Magritte, because this discussion may prepare the ground for understanding the role of frames in understanding a sign. Thanks to modern technology of reproduction, I believe that this picture must be familiar to most of the readers. Visually the painting is very simple. It involves a pipe and a sentence that appears below it: "Ceci n'est pas une pipe," or in English "This is not a pipe." The first time I encountered this picture I reacted with surprise: If it is not a pipe what is it?! A friend of mine who was smoking a pipe while I was looking at the picture looked at me quite amused and said: "Of course, that is not a pipe! *That is a picture of a pipe!*" It took me some time to realize that this painting is a wonderful occasion for inquiry into substantial epistemological problems. Indeed, it reminds us that the sign cannot be identified with the signified and that signs have meaning only within the dynamic of signs and meta-signs. Taking the components of this picture apart has its own intellectual significance, but its meaning is in its irreducible structure. As noted by Gablik (1982, p. 109):

> Magritte never dealt with single, static identities. His images incorporate a dialectical process, based on paradox, which corresponds to the unstable, and therefore indefinable, nature of the universe. Thesis and antithesis are selected in such a way as to produce a synthesis which involves a contradiction and actively suggests the paradoxical matrix from which all experience springs.

In this sense, Magritte's picture inquires into Being and exposes the fallacy of reification by questioning the referentiality of signs to real objects. The picture also involves a text ("This is not a pipe") which is an integral part of the whole visual representation it considers. What is the status of the text within this piece of art? If the truth-value of the pictorial representation ("This is a pipe" is truth) is negated by the text, what is the truth-value of the textual representation? Although the text seems to be secondary to the pictorial representation of the pipe, we may find out that the pictorial representation is secondary to the textual representation since the text seems to have authorship over the "pipe," authorship that cannot be disputed.

"Indeed," says the observer to the text, "It is not a pipe, but what are you? Who gave you the authority to decide what is a pipe and what is not? After all, this is not a pipe, but you are just a text. And if this is not a pipe, can I deduce that there is a real pipe in the outside world to which this representation refers?"

From a cultural developmental perspective, we may say that the first thing that existed was a pipe. A simple pipe with a concrete, material existence. Secondary to this material was the sign of the pipe, a representation of the pipe, painted on a piece of paper or canvas, and demarcated from the rest of the universe as a distinct sign. Then, at last, there was the whole sign that includes the text, "This is not a pipe." This is, of course, a simplified and inexact image, because I argue that at the basic form of existence there are *no objects but singularities that constitute the meaning of signs through some kind of mysterious dynamics*. In this sense, this is neither a pipe nor a representation of a pipe. In our case, the lesson is that by becoming semiotic beings very accustomed to the world of signs, we become ignorant of the dynamic nature of the universe and replace it with a static, definable world of objects. We should acknowledge the work of artists such as Magritte, who remind us through the alienation of familiar objects that those objects are not reality but some kind of a signifying activity. Is it? Or does Magritte's painting fall into the same pit it made us fall into? I believe that the painting and its lesson should not be taken for granted. The painting not only teaches us a lesson about people's ignorance of the relation between the signifier and the signified, but an important lesson about *the context of understanding the meaning of signs*. Let us discuss this conclusion further through the next dialogue with the cat.

Cat-logue 4
Where Does the Frame End?

Bamba: Hi, Dr. N What are you doing?
Dr. N: Oh, hi, Bamba., Come and see. I am looking at a most interesting painting by Rene Magritte.
Bamba: What is so interesting about this painting?
Dr. N: You see there is a pipe and at the bottom of the painting it says: "This is not a pipe."
Bamba: And this "This" refers to the above pipe or to the whole painting of which the text is a part?
Dr. N: That is an interesting question, but you see, the amusing effect is created by painting a pipe that flashes through our mind "pipe" and adding a text that reminds us that this is not a pipe, but a representation of a pipe.
Bamba: Ok, ok, I think I got it. What you are saying is that people are excited to find out that this is not a pipe but a *representation* of a *real* pipe.
Dr. N: This time you got it quickly.
Bamba: However, do you think people are so stupid as to mistake this representation for a real pipe? I mean... is there anyone who approached the pipe and tried to fill it with good tobacco in order to smoke it? Maybe a primitive man? A child? A cat?
Dr. N: Hmmm...I am not familiar with such an incidence.
Bamba: Or do you think that museum curators in the States would refuse to show this painting because "smoking is not allowed in the museum"? By the way, do you know that the British playwrite Dennis Potter said once that in the States it is easier to pull out a gun than to pull out a cigarette?
Dr. N: Please, stop it! You lack basic manners and I am starting to regret that I don't have an obedient dog rather than a sophist cat. At this moment, I am afraid that my American readers will throw this book out and you will be the ultimate cause of this diplomatic incident.
Bamba: You mean a decrease in the sales of the book?
Dr. N: Yes, I do.
Bamba: Less money?
Dr. N: Yes.
Bamba: Less superb food for me?
Dr. N: It seems that you understand almost everything very quickly as long as it concerns your belly.
Bamba: Recall Hegel: I am a concrete and primitive creature! Nevertheless, in this case, I humbly apologize. You see, what bothers me most is that if this painting is a representation of a pipe, then somewhere outside the painting there must be a real pipe.
Dr. N: And what is so disturbing about this idea?

Bamba: You see, the whole punch line of the painting is that we mistake the representation for the thing that exists out there in the outside world.
Dr. N: So?
Bamba: So...even in our first dialogue we reached the conclusion that "things" do not really exist out there in reality, but just singularities, interactions with singularities, that only on a second level of analysis, on a higher semiotic level of analysis, are things defined as demarcated entities from the organism which conceives them. I believe that you also reached an important conclusion that "things" exist only on the semiotic level and as "representations" or signs only. You also mentioned Volosinov and the idea that signs have meaning only in an interactional context, only as a response, and only as the result of micro-macro level dynamic between the sign and the whole utterance. At the bottom of those signs, you will not find the thing itself, but just processes of differentiation.
Dr. N: So, what you are suggesting is that this pictorial pipe does not correspond to some *thing*? To a real pipe?
Bamba: Indeed, this is neither a pipe nor a representation of a real pipe. It is the *invention* of a pipe through its signification. Therefore, I would like to suggest that the amusement human beings express when looking at the painting results from their reified *frame* of the universe. In fact, *the meaning of the picture is clarified when we realize that the boundaries of this picture are not the material boundaries of the material canvas but the boundaries you people force on your mind.*
Dr. N: Very interesting. In fact, I am a little bit bothered by the fact you have the best lines in the dialogues.
Bamba: Due to the fact that you are inventing these dialogues, it is really surprising that I have the best lines. As you can see, even in order to understand our conversation the boundaries of this dialogue should be expanded beyond its concrete frame to a meta-frame.
Dr. N: How about giving this cat-logue the title: "This is not a dialogue."
Bamba: So, what is it? I'm kidding!!! Let's go to smoke a pipe.

Chapter 9
A Snake that Bites its Tail

> Summary: The dynamic of structures and metastructures lies at the heart of any semiotic activity. This chapter explains this phenomenon and suggests acknowledging its importance for understanding the reification of the world.

The lesson taught by the cat in Cat-logue 4 concerns the importance of *frames* in understanding the meaning of a sign. But what is a frame and why is it so important for understanding the meaning of a sign? In one of his seminal papers, "A Theory of Play and Fantasy," Gregory Bateson (2000) presents the idea that living communication systems operate at several levels of abstraction, and he differentiates between metalinguistic levels of abstraction and metacommunicative levels of abstraction. Metacommunicative messages are messages where the subject is the relationship between the speakers. For example, after telling a joke that could have been interpreted as an insult, one may say, "I was just joking." The metalinguistic level of abstraction involves messages where the subject is the language. For example, the utterance "the word 'cat' is not a cat" is a metalinguistic message that says something about the meaning of the word "cat" and implicitly about the status of words in general. People who have difficulties in moving between levels of abstraction and grasping metalinguistic messages may confuse the map with the territory, sense with reference, or the sign with the signified. In fact, by living in a reified universe it seems that all of us confuse the signs we use with real "objects." There is a nice fable, illustrating this difficulty, about a man who comes to a new town and desperately looks for a place where he can press his trousers. After anxiously looking for such a place, he sees above a store a signboard "We press trousers here!" The man happily runs into the store and to his disappointment finds out that this is a store that manufactures signboards!

The importance of meta-messages was evident to Bateson when he observed young monkeys playing at the San Francisco zoo. The interaction between the monkeys looked like a combat but it was not (This is not a fight = This is not a pipe) and the monkeys seemed to be well aware of it:

> It was evident, even to the human observer, that the sequence as a whole was not combat, and evident to the human observer that to the participant monkeys this was "not a combat." Now this phenomenon, play, could only occur if the participant organisms were capable of some degree of metacommunication, i.e., of exchanging signals which would carry the message "this is play." (Bateson, 2000, p. 179)

The monkeys play by pretending to fight. They exchange messages of fighting, but they were not fighting. They exchanged messages that were untrue or not meant in the sense that they denote something that does not exist (a non-existent pipe). This semiotic activity, like the exchange of money, is possible only by the meta-frame that, on the one hand, allows the existence of those messages and, on the other hand, restricts their referential meaning. That meaning of a sign is determined "in between" the dynamic object (or the activity) it somehow is supposed to signify and the meta-frame that constitutes its meaning by qualifying its referential meaning. Meaning is constructed through the *paradoxical movement* between the referential meaning of the sign (This is a pipe) and the frame that negates it (This is not a pipe). Unfortunately, from time to time our meta-frame ceases to function and the result is that pretend combat becomes real. This phenomenon is evident in children's play (and worse in adult life and philosophers' metaphysics) when play combat turns into a genuine and painful fight.

The existence of metacommunicative frames, but in a more general sense, is also evident when we watch a movie. Some of us enjoy seeing horror movies in which scarey Hollywood monsters feast on innocent citizens from a peaceful rural town somewhere in the American Midwest. The people who watch the movie enjoy it as long as their metacommunicative frame tells them "those are not real monsters." More accurately, they enjoy the "in between" of the semiotic event (the movie) since if the monsters are real monsters, their enjoyment would immediately cease. On the other hand, if they are fully aware that those creatures on the screen are not real monsters, what is the point of watching the movie and enjoying the thrill? This is, of course, a delicate balance. When my older daughter watches Peter Pan, she is both fascinated and terrified by the frightening Captain Hook. At certain moments, her fear overcomes her meta-frame and she runs to me anxiously. At those moments, my parental role is to remind her to use her frame by saying to her "Don't be afraid, this is *just* a movie," as if the *just* is some kind of an Ontological Safety Valve prohibiting our signification process from falling into the real. When the Lumier brothers presented the first movie, the French audience escaped from the theater, scared to death that the train they saw on the screen would run them over. This audience did not have the appropriate frame for watching a movie and, therefore, interpreted the message literally. How is the relation sign-frame relevant to our analysis?

As will be argued later, the sign, by its internal dynamic, always turns in on itself to create a frame through which its existence may be constituted. That is, the sign and the whole of which it is a part each support the existence of the other in a spiral like process. This is why there is no language without metalanguage and vice versa. Pointing at the circularity of the semiotic process may remind us again what Volosinov has taught us: The semiotic event is always dynamic. This lesson presents us with a quandary. How is it that the semiotic event is always dynamic, but that our world, reflected in our semiotic mind, appears reified? Previously I pointed at the dynamic in which the circulation of the sign creates a stable appearance of value. This is the key for understanding the reification of the world. The dynamic of this process may be such that *the value of the sign appears to be determined by its reference to a real object (rather than by its being captive in a circular process) by oppressing its indeterminance and its dependence on the interpretative frame.* By disconnecting the sign from its context the world may appear reified. A world populated by signs that correspond to real objects, rather than a world in which our process of signification is the source of mind-reality. *The interesting thing is that this process is possible since the frame/context/metalanguage are usually hidden from us.* This invisibility is especially evident in art. As Duro (1996) suggests: "The task of any discussion of frames and framing in the visual arts is first and foremost to counter the tendency of the frame to invisibility with respect to the artwork" (p. 1). This suggestion follows Derrida (1987), who poetically points (at one of his less obscure pieces) at the *invisibility* of the frame:

> The *parergon* stands out both from the *ergon* (the work) and from the milieu, it stands out first of all like a figure on a ground. But it does not stand out like the work. The latter also stands out against a ground. The parergonal frame stands out against two grounds, but with respect to each of these two grounds, it merges into the other. With respect to the work which can serve it as a ground, it merges into the wall, and then gradually, into general text. With respect to the ground, which is the general text, it merges into the work, which stands out against the general ground. There is always a form, on the ground, but the parergon is a form which has its traditional determination not that it stands out but that it disappears, buries itself, effaces itself, melts away at the moment it deploys its great energy. (p. 61)

The lesson that we can learn from the above analysis is that the sign never exists outside the dynamic of a frame/context/metastructure that gives it meaning. There is no text without context although many cultural establishments would like us to believe so. If there is text without context, then we can understand the text "literally." There is one interpretation of it and this interpretation is the true one. Those who hold the right interpretation are the good guys. Those who hold the false interpretation are the bad guys. With regards to our quest we must conclude that the sign by itself could not have been transformed from a concrete form of existence (e.g., "This is a pipe") to an abstract form of existence (e.g., "This is not a pipe, but a sign of a pipe"). The dynamic of a sign and certain interpretative frameworks are those that should explain the reification of the world.

9.1. THE POETIC OF DETACHMENT

A sign by itself is meaningless. The sign by its nature is always oriented toward its dynamic object through which it achieves its value; This is a pipe! Here it is over there. Just search and you will find it. Since the value of the sign is derived from its relation to the referent (which is of a derivative value too: This is a pipe because I can signify it as such) it can never be identified with itself. The law of identity A = A is actually meaningless within this system. *Therefore, the identity of the sign, as materialized by its value, is a wraith that cannot be grasped either from within (the sign system) or from the outside world. Its presence must be constructed through the process of exchange and circulation without obeying the law of identity in its essentialist sense.*

This process of signification is of a mythological nature of power and threat. Man is aware of the enormous potential of his sign system to change his position in the world. From pre-objective to post-objective form of being-in-the-world. The sign system is an emancipatory system of being-in-the-world that constitutes the mind as a semiotic system and at the same moment detaches it from its embodiment in present time, matter and operation. However, this semiotic power is threatening. It is associated in the great mythologies with the God(s), the dawn of the human race (Hermes or Adam and Eve) and with the price that should be paid by human beings for its use. The power of symbolization is too demanding, ontologically absorbing, forcing us to look with open eyes into the heart of darkness, into the lacuna of our semiotic being, to the semantic stability of our value system, to the flux of being. No real ground for the words. No essence for things. No origins. No identity. A system of wraiths in a material world. Powerful wraiths associated with God, power, pain and action.

For solving the agony of its detachment-from-the-object, there are occasions (cultural, historical or even personal/temporal) in which the symbol is turned in upon itself as a meta-frame, to establish a language of communication, a reflective language of alchemy (that is, a scientific language of prescriptions) that does not seek to transform the metal into gold, the referent into the symbol, but the gold into gold by establishing the symbol as an ideal type of being with an identity preceding its value. A process in which the symbol turns upon itself as an amnesic suppression of its epistemological position, of its ontological orphanhood. This process is evident in all of the meta-frames that separate the mind from the world. In this world and by becoming the thing-in-itself the symbol is not the abstract entity we thought it was. It does not have any relation to its mythological-processual origin, but the one of value only. It is not the abstract any more, but the imagined, the detached, the glorious, the pure and therefore the authoritative that determines the (quasi) existence of the referent as subordinate to it. This framework saves the sign from the lacuna of being, from the non-existence of a stable signified, by inventing itself as reality.

The turn of the sign inward is expressed in a movement known as "symbolism," as the return to the self/soul as the source of poetic creativity, in contrast with the naturalist/mimetic vision in which the sign turns outward. Therefore, the abstract movement in art is a concrete expression of the Platonic framework of appearance vs. reality, soul vs. matter, and memory vs. senses as a gate to reality.[31] It reminds us of Plato's famous cave allegory:

> And now, I said, let me show in a figure how far our nature is enlightened or unenlightened: Behold! human beings housed in an underground cave, which has a long entrance open towards the light and as wide as the interior of the cave; here they have been from their childhood and have their legs and necks chained, so that they cannot move and can only see before them, being prevented by the chains from turning round their heads. Above and behind them a fire is blazing at a distance, and between the fire and the prisoners there is a raised way, like the screen which marionette players have in front of them, over which they show the puppets.
> I see.
> And do you see, I said, men passing along the wall carrying all sorts of vessels, and statues and figures of animals made of wood and stone and various materials, which appear over the wall...
> You have shown me a strange image, and they are strange prisoners.
> Like ourselves, I replied. (Plato, 1970, p. 296)

9.2. GETTING OUT OF THE CAVE OR SOLVING A RIDDLE?

For Plato the only way out of the cave is an *in-sight* into our soul, an insight that aims to uncover the abstract, firm and genuine nature of reality. However, there is another tradition of inquiry and frame of mind that emphasizes the *dynamics* of reality and our ability to reveal its existence not by removing the screen of appearances but by rearranging the riddle of appearances in a logical way that corresponds to the latent structure of the cosmos. This frame is epitomized by Heraclitus (500 BC). Due to the fact that Heraclitus left only mysterious short fragments, it is quite difficult to argue with complete confidence that he had a coherent philosophical theory. However, there are scholars who assert that this is possible. Even if we consider this scholarly speculation, we should admit that it is an interesting one that should be examined seriously.

For Plato, Heraclitus is the "theorist of a universal flux (*panta rhei*)" (p. 4). Hussey (1999) suggests that the famous fragment attributed to Heraclitus, "you could not step twice into the same river," is non-Heraclitean. However, following Plato and Aristotle, Hussey suggests that for Heraclitus, *process* is "the basic form of existence in the observable world, although something, not directly observable, persists throughout" (p. 99).

Like many philosophers, Heraclitus is aware that the senses cannot provide us with a clear insight into reality since they are already shaped by our preconceptions: "Eyes and ears are poor witnesses for men if their souls do not understand the language." Therefore, "the possibility of understanding is correlated with the existence of a meaning. It implies the need for interpretation of what is given in experience, as though it were a riddle or an oracle" (Hussey, 1999, pp. 90-91). According to this interpretation, reality is hidden. It is a hidden structure, and there is no algorithmic way to uncover it. However, once the solution to the riddle is found, there is no doubt that this is the solution. The same is true for the riddle of appearance. We should trust the solution that appeals to the logos – to the reason – as manifested by the agreement of a shared community.

Surprising as it may look, Heraclitus presents a social theory of epistemology. Not that he denies that there is a "true" structure underlying the appearances. However, he suggests that since there is no algorithm that may lead us beyond the screen of appearance we should count only on the indirect way, and to confirm our findings through some kind of social agreement. This is a farreaching statement for a scholar and it may totally destroys his/her academic credibility! That the logos of the universe should be uncovered by social agreement? This sounds peculiar, but the argument is not that any statement regarding the structure of the universe should be based on some kind of social fashion, but that the heart of any scientific activity is *interpretation,* which is a *social activity,* and the result of this activity is so clear that it is evident to the public and cannot escape social recognition.

Concerning Man, Kahn (1981) argues that Heraclitus was mainly interested in "a mediation on human life and human destiny in the context of biological death" (p. 21) and that "in Heraclitus' view such understanding of the human condition is inseparable from an insight into the unifying structure of the universe, the total unity within which all opposing principles – including mortality and immortality – are reconciled" (p. 21). Therefore, we can say that for Heraclitus Man is a micro-cosmos and that people's failure to reveal this truth results from their failure "to comprehend the *logos* in which this insight is articulated, the *logos* which is at once the discourse of Heraclitus, the nature of language itself, the structure of the psyche and the universal principal in accordance with which all things come to pass" (p. 21-22). As Heraclitus writes: "You will not find out the limits of the soul (psyche) by going, even if you travel over every way, so deep is its report" (XXXV). As interpreted by Kahn: "By seeking for his own self, Heraclitus could find the identity of the universe, for the logos of the soul goes so deep that it coincides with the logos that structures everything in the world. Hence, the error of those men who treat thinking as private, in the face of the fact that "the logos is common" – common to them and to everything else" (p.

130). Bravo!!! Opening our quest with Husserl's striving for the essence through the phenomena, we get to Heraclitus who suggests quite a similar idea. However, transcending appearances or using them as a springboard for uncovering the structure of reality are methods that seem to avoid a direct confrontation with the problem of the dynamic of structures and metastructures and its spiral like nature. Is there another way in which our inquiry can proceed? A logic that takes the dynamic of structures and metastructures as its point of departure?

9.3. IS GOD A POST-MODERNIST?

The Talmudic case I would like to present[32] in this section concerns a dispute among several scholars regarding a question of whether a specific kind of oven is susceptible or unsusceptible of being restored to its ritually pure state. That is, whether a specific kind of oven is subject to the Jewish laws of defilement or whether it is an oven that cannot be rendered impure and therefore the laws of defilement do not apply. This question seems a little bit strange to the modern secular mind, but those scholars were engrossed by issues of impurity, not in the hygienic sense, but in the theological sense, and this issue was an integral part of their world. The question at stake is not the most important issue in the Talmudic story, but rather the drama that evolves out of this specific dispute. At the beginning we are told that the leading sage, Rabbi Eliezer, declares that the oven may be broken down [because it is made out of discrete units] and therefore is not susceptible of becoming ritually impure. In contrast, the other sages declare it is susceptible.

> On that day Rabbi Eliezer produced all the arguments in the world, but they did not accept them from him. So he said to them, "If the law accords with my position, this carob tree will prove it." The carob was uprooted from its place by a hundred cubits - and some say, four hundred cubits.

What we see is that after trying to convince the other sages in a rational way by providing arguments in favor of his position, Rabbi Eliezer gave up the rational way and tried to convince the sages by using supernatural corroboration. Surprisingly, the carob tree was uprooted from its place. Does this miracle convince the stubborn sages? No. And they react to this miracle by saying, "There is no proof from a carob tree." Rabbi Eliezer was a leading sage and a descendent of Moses himself. A sage not to be easily dismissed. The response he received from the sages is therefore unacceptable. So he tried another miracle and said: "If the law accords with my position, let the stream of water prove it. The stream of water reversed its flow." At this point of the story, the reader is assured that the sages must be convinced by Rabbi Eliezer's position. One miracle, such as a moving carob tree, may be explained by chance ($p < .001$!), but a stream of water that reverses its flow?! This is really something that should be seriously considered. The stubborn sages seem to acknowledge this miracle but they refuse to accept it as a legitimate source of evidence by saying: "There is no proof from a stream of water." So Rabbi Eliezer tried another miracle and again failed to convince the sages. At this point, it seems that he became very angry, frustrated and turned for the assistance of the Almighty:

> "If the law accords with my position, let the Heaven prove it!" An echo came forth, saying, "What business have you with Rabbi Eliezer, for the law accords with his position under all circumstances!"

Aha! This time the old Rabbi Eliezer seems to be succeeding. Receiving support for your position from God himself is something no one can argue with unless… he is a Talmudic sage. One of the sages, Rabbi Joshua, stood up on his feet and said, "It is not in heaven." As interpreted by Rabbi Jeremiah, the meaning of Rabbi Joshua's interpretation is that:

> "Since the Torah has already been given from Mount Sinai, so we do not pay attention to echoes, since you [God] have already written in the Torah at Mount Sinai, 'After the majority you are to incline.'"

The story is close to its climax, since the Talmudic sage is arguing with God himself by reminding him that the majority [of sages] has the authority to decide on matters of law. Is this Talmudic sage rebelling against God himself? How should God react to this rebel? Will he punish the rebel sage? Will he stab him with the carob tree? At this point, I feel like a viewer of a soap opera and I have a strong urge to say something like, "The answers to these questions and more in the next chapter of the Talmudic Saga!" However, I assume that the readers are too curious to wait and see how God responds. The same curiosity pushed Rabbi Nathan to ask Elijah [the dead prophet who suddenly get into the story]:

> "What did the Holy One, blessed be he, do at that moment?" and he [Elijah] replied to him: *"He laughed and said 'My children have overcome me, my children have overcome me!'"*

This is an amazing story. A group of sages is arguing about a specific, even technical matter. Although one of them seems to be in the right, supported by God himself, the sages refuse to accept his position and argue against God that he himself commanded them to adhere to rules of interpretation and scholarly consensus and not to rely on miracles. This position sounds paradoxical, because the same God who created the law is evident at this specific dispute, and clearly supports one of the sages. Therefore, what is the point of reminding God that he obliged them to do something else in the past? Does he not remember it by himself? Is the will of God subordinate to the interpretative rules he determined? And why does God laugh and say, "My children have overcome me"? Does

God laugh? Can he be overcome through social consensus and rhetoric? Is he a relativist who changes his laws according to social norms? Or in other words, *is God a Post-Modernist*? We must admit that this story is very strange.

My reading of the Talmudic story takes us back to Plato, Heraclitus and the appropriate way of understanding mind-reality. Although both Heraclitus and Plato present the idea of a reality as differentiated from appearance, it seems that there is a big difference in their positions. For Plato, the world of appearance is a mask that covers a hidden truth. In order to uncover the truth one has to tear off this mask by overcoming sensual data and introspection [interospecting?] into the secrets of reality. In modern hermeneutic terms, we can say that the text covers the truth and one should *transcend* it in order to grasp its meaning. In contrast, my reading of Heraclitus suggests that the "text" should not be neglected in order to reveal the latent structure of reality. If the text is like a riddle, it should be carefully interpreted in order to organize its internal structure and by that to uncover the logos which is reflected in it. This conception suggests that there is no difference between "reality" (*meaning*) and "appearance" (*text*) as there is no difference between *nomos* (convention) and *physis* (nature), or between knowledge of the self and knowledge of the cosmic order. These distinctions are later conceptions that Heraclitus did not hold with. In other words, Heraclitus presents a theory in which the general order of the cosmos (reality) is reflected, but maybe in an enigmatic way, in the world of appearances. This position seems quite similar to the hermeneutic position governing the Talmud. However, the Talmudic story takes us in another direction, far away from the epistemic story as told by Plato and Heraclitus.

In general, the Talmudic inquiry, as evident in the Talmudic writings, is not interested in *epistemology* and *aesthetic;* those domains have deep connection, since both are nurtured from the tension between appearance and reality or from the tensions between contradicting appearances. Talmudic inquiry is about *praxis* and *ethics*,[33] about the way the divine will should be realized in man's practice in the world. The difference between the Greek conception of search for order and the Talmudic conception of search for order is qualitatively different. The Talmudic inquiry does not accept as its epistemological starting point a gap between logos and its representation, between reality and appearance, but adheres to a position in which man's practice in this world, his unique way of being-in-the-world as a Jew, is the logos.[34]

Let me try to be more explicit concerning the ethical and practical Talmudic method of inquiry. According to the above interpretation of Talmudic method, logos is not masked by appearance, nor is it a riddle that needs solving. It exists neither in the text (appearance) nor in some kind of a Platonic realm, but at the *dynamic event of interpretation which is social through and through*. The activity, the practice (in this case the activity of interpretation), is what constitutes logos, which in turn constitutes the meaning of the interpretative activity. In this sense, the basic unit of reality is neither the thing-in-itself nor the Cartesian self, but a *Holon* (wholes within themselves while at the same time a part of a larger whole, and the other way around) *of interpretative activity* in which the whole activity provides the meaning of its units, and the units provide the meaning of the whole in an undifferentiated and ongoing process. In the Talmudic case, the sages' decision is based on a scholarly consensus that receives its authority from the interpretative framework the Lord instructed the sages to use. The validity of the interpretation is derived from a given interpretative framework (the whole). At the same time, the interpretative framework receives its power from a group of people (the sages) that interpret it as their interpretative framework! This is the point where the notion of the Hermeneutic circle, the *ourobouros*[35] (the legendary snake that bites its own tail) and many other forms of circularity and recursion come to mind. The structure of mind-reality is constituted by a self-referential interpretative dynamic that derives its meaning from a spiral process of signs and meta-signs that guide us, whether we are "lower-order" organisms that interpret simple signs or "superior human beings" who use symbols to communicate with one another. Wisdom involves the acknowledgment that there is no reality-in-itself behind the curtain of our mind, neither is it the mere appearances of our mind. Logos exists at the action and at the process by which mind is constituted as a self-referential interpretative activity significant only with regards to a concrete communicational context, as suggested by Volosinov, Bateson and the stubborn Talmudic sages.

In the next part of the book, I will try to develop our understanding concerning the dynamic of signs and meta-signs, and the way they transform us from our primordial position of being in the world to a reified universe governed by signs pretending to be the thing-in-itself.

In sum, we opened the first part of the book by recognizing that we reflect on our universe as a reified universe, a fragmented universe composed of objects. We questioned this appearance and argued that at the most basic level of existence this appearance does not seem to reflect the real state of affairs. The contrast between what appears and what is there has motivated us to inquire into the content of our consciousness acts. We found out that our mind is mediated by signs and that we should better understand the meaning of our semiotic mediation in order to understand the reification of the world. After surveying the developmental and the non-developmental conception of the sign, we realized that the social is an essential term in understanding semiotics. We turned to Saussure to adopt the idea that we should study semiotics as a social-systemic phenomenon and proceeded to Volosinov to find out that the mind is a semiotic interface constituting meaning through the dynamic between the sign and its frame in a local context of communication between people. Our next step was to inquire into the cultural development of the sign system and to reject the simple developmental

A Snake that Bites its Tail

theory of Hegel and his disciples in favor of the idea of the dynamic of structures and metastructures that constitutes the meaning of a sign. I presented the idea that the sign out of its internal dynamic generates meta-frames that on certain occasions may suppress its dynamic origins and result in an appearance of value which is allegedly derived from a reference to a real object. We concluded by pointing at the self-referential dynamic of our mind as the most basic unit of our inquiry and suggested inquiry into the origins of mind and the reification of the world by taking this *Holon* as our most basic unit of analysis. Let us proceed to the next part and reconstruct mind from its origins by adopting the conclusions we drew from the first part of the book.

Chapter 10
The Demon of Circularity

> Summary: The world as it appears to us is a world inhabited by objects. If we are an object among these objects, this is a situation that presents enormous difficulties for self-knowledge: How can we establish firm foundations for knowledge in a world of which we are a part? A possible solution to this difficulty is to consider self-referential activity as the basic unit of existence, rather than as a well-defined object (the "I").

In the previous chapters, I presented the idea that the origins of mind are embedded in a primordial and processual experience with reality. In this primordial domain, the law of identity does not hold, and being, in the purest sense of the term, is a relational structure with the world. I argued further that only by entering the realm of the sign do we fixate the world and reduce it to small, static pieces. Moreover, by using these signs and their framework, we reflect on the world and theorize about it as a reified universe. By shifting from a non-symbolic primordial (or a pre-symbolic) form of existence to a symbolic form of existence, we have obligated ourselves to a system of signs that have meaning as long as they are circulated among the members of a community, and that the ultimate source of fixation has something to do with our socio-somatic-semiotic form of being and interpretative frames. In the following chapters, I elaborate on these ideas and present the argument that the *reification of the world is evident when we ignore the fact that our relation with the world is both through reason (mediated signs) and fantasy (the belief in its stability), and let reason overcome fantasy*. The current chapter opens this inquiry by examining the possibility of "knowing" our mind in a world in which knowing is a spiral-like process. This question is a part of a long historical attempt to understand the nature of knowing.

The term "epistemology" originates from Greek and is composed of *"episteme"* (knowledge) and *"logos"* (explanation). In general, epistemology deals with the meaning of knowledge, the conditions of knowledge and the limits of knowledge. Since this topic is widely explored in basic philosophical textbooks, I do not elaborate the theoretical and the historical foundations of epistemology further. However, I do want to stress the main drive of epistemological research across time and disciplines, which is the drive to establish knowledge on solid ground. This by itself is not a unique property of epistemology. The activity we usually describe as "science" in its different manifestations also strives to establish itself on solid ground by using axiomatic systems, logical reasoning, empirical evidence and so on. So what is unique about the ideal of epistemology to establish itself on solid ground?

The uniqueness of this venture is derived from a basic difficulty facing epistemological research: Trying to achieve a solid base of certainty concerning the way we know the world of which we are a part is a recursive process. This in turn is extremely difficult to grasp through traditional methodologies of inquiry familiar to us. In other words, and in contrast with other scientific domains, in epistemology the observer is observing himself, trying to understand the process of understanding. In contrast, in other scientific domains we usually conceive the object of our inquiry (atoms, genes, cells, animals, etc.) as qualitatively distinct from the observer (e.g., the scientist). This is the major difference between what is known as the two cultures: The culture of natural and exact science, in which the difference between the subject and the object does not usually concern the researcher, and the culture of the humanities and the social sciences, in which the subject attempts to know himself, and the way (s)he knows the world constitutes *the* difficulty facing the researcher.

In order to illustrate the difficulty embedded in the epistemological venture it may be helpful to use one of the tales associated with the famous Baron von Münchhausen. The Baron, one of my childhood heroes, is a legendary figure who finds himself involved in many dangerous adventures. The Baron attained his reputation by getting himself out of trouble in creative ways. In one of his journeys, the Baron was riding his horse and they both fell into a muddy swamp. This situation was life-threatening. The horse was sucked down into the swamp and the Baron, who was sitting on its back, could not get the horse out. Think about yourself. What would you have done in a similar situation? As always, the Baron found a creative solution to his problem. He wrapped his legs around the horse, grabbed his own hair, pulled it *up* and hoisted himself out of the swamp. Kids, do not try to do this at home! It is just a story that illustrates the conundrum of a reflective act and the illogical image of the bootstrapping phenomenon.

We realize that the Baron could not hoist himself out of the swamp. This is what makes the story an amusing tale. The epistemological analogy is that one cannot really know the way one knows by stretching one's *individual* brain further. This possibility seems to us fantastic and similar to the legendary snake that bites its own tail.

A conundrum.

A paradox.

It seems that in order to solve the mystery of recursive knowledge, we should look for an *Archimedes point* of epistemology, a point outside or inside the observer, through which the whole structure of human consciousness can be constituted. Is there such a point? Western culture is still governed by two grand narratives – the rationalist and the empiricist – that have attempted to locate the *Archimedes point* either outside the subject (the empiricists) or inside the subject (the rationalists). In this book, I present a rival suggestion that the *Archimedes point* should

The Demon of Circularity

be sought neither inside nor outside the subject, but *in between,* as suggested by Merleau-Ponty and others.

Let us recall one of the most impressive attempts to constitute consciousness from within. Rene Descartes was one of those who sought an Archimedes point *within* the observer. Descartes was a skeptical philosopher in the sense that, in order to achieve a solid base for knowledge, he did not want to accept any belief as valid just because of societal norms or other suspicious sources of belief. He even questioned the existence of the outside world as appearance that could have been created, like a dream, by a vicious demon. In contrast with Peirce, Descartes was highly suspicious of *any* form of mediation between the inquiring mind and truth. In this sense, we may describe Descartes, although he was a Catholic, as an "epistemological Protestant" who refused to accept any form of mediation between mind and reality, between God and Man.

Descartes adopted a position we may describe as "tactical skepticism" and doubted the existence of everything in order to establish a solid foundation of knowledge. Can one trust the "outside" world when all appearances can be the satanic invention of a vicious demon making fun of our senses?

In an extreme phase of his inquiry, Descartes even doubted his own existence. This extreme doubt lead Descartes to the first truth of the Cartesian quest: The existence of the thinking subject. Precisely because I can doubt my existence I must infer the existence of a creature who is able to doubt. I think therefore I exist – *Cogito ergo sum.* This is an important phase in the history of Western philosophy. There are good reasons to suspect that this argument is totally wrong, but the intellectual efforts should be praised.

From the phase of establishing an Archimedes point of certainty – the immaterial "I" – Descartes goes on to prove the existence of God as a perfect being and, through the existence of the Almighty and his excellent character, to prove the existence of the "outside" world. Unfortunately, Descartes cannot avoid falling into circularity: If the reliability of the intellect depends on our knowledge of God, how can knowledge be established at first place?!

Being afraid of a demon that might obscure his mind Descartes finally found himself struggling with the most vicious demons of all – the demon of circularity.

From the discussions presented in the previous chapters, we know that circularity is an essential aspect of living beings. There are no living creatures that do not express circularity in their behavior. Therefore, trying to dismiss circularity and to avoid it while theorizing about the mind seems hopeless. I believe that in a way similar to the mythological monster – the Hydra – the demon of circularity can be defeated or, better, domesticated, by disconnecting it from its specific ontological grounds. Demons tend to flourish well only on specific ontological soil, and the demon of circularity seems to flourish in a world composed of objects. Therefore, I suggest domesticating the demon of circularity by disconnecting it from our reified ontology. The next chapters present the idea that the mind is a self-referential process that emerges from reality through a socio-somatic-semiotic process of differentiation.

Chapter 11
Origins

> *Our view of man will remain superficial so long as we fail to go back to that origin, so long as we fail to find, beneath the chatter of words, the primordial silence, and as long as we do not describe the action which breaks this silence.*
> **Merleau-Ponty**

Summary: The basic unit of the mind, and of any form of living being, is a process of differentiation. A process in which a distinction is made within a void. This process appears in mystical texts and suggests that nothingness is the only reality we really know and that all appearances result from this reality through a process of differentiation.

In order to identify the most basic unit of our mind, we should look for the most basic unit of mental activity, the unit that differentiates the emergence of *something* out of *nothingness,* in other words, to identify the "Big Bang" of the mind. It must be noted that I do not aim to trace the historic origins of the mind, but to inquire into its metaphysical origins.

The thing about nothingness is that it is such an obscure concept that even the English language can really handle it only through a negation of something that already exists (no-thing). I want to elaborate further on the concept of "nothingness," but I have nothing to say about it. I hope it has nothing to do with my intellectual abilities, and if this deficiency in saying something about nothingness has something to do with my intellectual abilities, I want to know nothing about it. Simply nothing!

I do not believe that my silence in the face of nothingness has something to do with my limited intellectual abilities, because nothingness cannot be characterized in positive terms: "It has no definition, no differentiation, no distinction. When all is the same, when all is one, there is no-thing, nothing" (Robertson, 1999, p. 46).

Although nothingness is a mysterious concept, it is a concept with a primary importance because, both in terms of ontology (what there is) and epistemology (how we know), it is the first concept that pops into our discussion; something (being) must emerge out of nothingness unless you adopt an Aristotelian position that regresses from what there is to what was before it ad infinitum.

The concept of nothingness (or non-being) is central to most mysticism. Although mysticism is sometimes associated with naive New Age ideologies, with young blurry-eyed teenagers with a limited knowledge of physics who speak about quantum mechanics and Gaia at the same time, or with some kind of holistic thinking that connects anything to everything and vice versa, do not let this image mislead you. To recall one of Nietzsche's insightful aphorisms, the followers of a system cannot be considered evidence against the system. Indeed, mystical thought is a courageous and dangerous attempt to touch and transcend the boundaries of our mind. In this context, the concept of nothingness fulfils an important role, and for some mystics such as Dionisus the Areopagite and the Kabbalist David Ben-Avraham, nothingness is equal to God. It is interesting that this idea also appears in Eastern philosophy. For example, Lao Tzu says: "The Tao that can be told is not the eternal Tao. The name that can be named is not the eternal name. This nameless is the beginning of heaven and earth."

The idea that something emerges out of nothingness (non-semiotic = nameless reality) is known as the doctrine of creation *ex nihilo*. This doctrine was not Jewish, but got into Judaism via Christian and Muslim thinkers under the heavy influence of Plotinus. Interesting discussions concerning creation ex nihilo exist in the Kabbalistic texts and the transition from nothingness to something is one of the most disturbing problems those texts try to solve. In this sense, theoretical biologists who try to trace the origins of life to inorganic (dead?) matter follow the footsteps of those mystical thinkers, and in many cases with a similar level of theoretical clarity and success.

The words, the signs we use, play a crucial role in this differentiation process. In the Book of Genesis, we recall, the world was created by words. This idea echoes in statements such as the following one by the most famous mystical thinker of the Middle Ages, Meister Echart, who said: "If God stopped saying his word, but for an instant even, heaven and earth would disappear." This is, of course, an interesting idea. However, we should also recall that the name we give to things, the distinctions we made, exist (a la Saussure) as long as they are differentiated. The world of signs (the *indications* we give to *distinctions*) can be reduced to distinctions. Can we reduce the distinction? Yes. To nothingness. Therefore, *a distinction is the most basic irreducible structure we have.* Beyond this realm, the realm of distinctions (appearances) and their indications (signs), there is only mystery. As Nagarjuna says: "The ultimate truth transcends all definitions and descriptions, transcends all comments and disputations, transcends all words." Those who try to study the way something emerges out of nothingness may find themselves challenging their mind with what is by definition beyond the mind as a "distinction machine."

I believe that readers who encounter mystical texts are caressed by the obscurity and the challenging ways in which those texts try to resolve the difficulties underlying creation ex nihilo. Some of us may feel quite uneasy with those attempts. Is there a way that realizes the emergence of something out of nothingness and at the same time avoids the obscurity of mystical texts? I believe that there is such a way that preserves the sense of mystery associated with the "Big Bang" of existence, but at the same time commits itself to a clear logic of thought. In order to expose this way, let us meet the stubborn Talmudic sages again.

The sages of the Talmud recognized long ago that the limits of our understanding lie at the intersection of nothingness and the things that emerge out of nothingness. This recognition is illustrated in a discussion between two Talmudic sages concerning the reason that the divine act of creation, as described in the Book of Genesis, begins with the Hebrew letter "bet." Bet is the second letter on the Hebrew alphabet. It is the first letter of the first word that appears in the first chapter of the Bible. This letter is read *from right to left* and looks like that:

The discussion appears in a text we previously mentioned: Breshit Rabba:

> R Jonah said in R Levi's name: "Why was the world created with bet? Just as the bet is closed at the sides but open in front, so you are not permitted to investigate what is above and what is below, what is before and what is behind."

This interpretation suggests that analogically to the "bet," which is spatially closed at the sides and opens at the front, our knowledge is constrained by the point where something emerged out of nothingness; beyond this, there is only mystery and there is literally nothing to say about it.

This position states that the primary source of existence, the emergence of something out of nothingness, is a mystery and that should be recognized and adopted as our point of departure. Any discussion (a serious one of course) is both constrained and constructed by this mystery.

There is, however, a contradicting point of view presented by another sage, Henry Bergson, who considers the emergence of being out of non-being as a false problem. Bergson, who we will also meet again later in this book, argues that many of the philosophical problems that have bothered Western thinkers are in fact false or nonexistent problems. Simply stated, his argument is that there is more in non-being than in being since non-being assumes being plus its negation. Therefore, inquiry into the emergence of being from non-being, or asking why there is something instead of nothing, is false since we behave as if non-being precedes being and not the other way around. This argument cannot be easily dismissed as long as one considers non-being as ontologically inferior to being. In other words, if we consider non-being to be the negation of being, then Bergson's argument is convincing. However, we should recall that our definition of non-being is totally different and that the use of the term "non-being" is a shortcoming of the human language rather than a deliberate and successful philosophical choice. Language, although a system able of meta-use, comes up short of providing us with the best means to transcend the last boundary of the mind.

Cat-logue 5
The Hole in the Bagel

Bamba: My impression is that the concept of nothingness has been underestimated in modern Western culture.
Dr. N: Yes, I can understand that. Since nothingness does not exist, it does not seem to have any relevance to us.
Bamba: This is a biased point of view because in practice nothingness or non-existence have much more relevance to us than "existing" things.
Dr. N: You must have gone out of your mind!!! How is it possible that non-existence is relevant to us?
Bamba: Do you know the joke about a man who comes to a deli and orders a bagel?
Dr. N: No.
Bamba: So this person comes to a deli, orders a bagel, and after few minutes the waiter comes, serve him an empty plate an says: "Sir, we have run out of bagels, and we were left only with the holes."
Dr. N: Very funny. I did not know that cats have a sense of humor.
Bamba: Why are you so prejudiced?! Haven't you read Alice's Adventures in Wonderland? Don't you know that the non-human has also a sense of humor?
Dr.N: In any case, what has this joke to do with nothingness?
Bamba: Does a hole exist?
Dr. N: Definitely not. It is an empty space, a parasite of the bagel.
Bamba: This is the point. You see, according to your perspective, the hole does not really exist. However, the hole defines the bagel. There is no bagel without a hole.
Dr. N: Let me try to understand. You are saying non-existence is secondary to existence?
Bamba: On the contrary! Existence is secondary to non-existence.
Dr. N: How is it possible? Are you suggesting that there can be a hole without a bagel that surrounds it?
Bamba: Definitely! It exists everywhere. I mean nothingness. Something comes into being only by differentiating a void. And, by the way, think about the holes in your body. Can you exist without them? You could try it once, but that might be a painful experience. On the other hand, I am sure that the nothingness, which is demarcated by your body, can rest peacefully without your existence. Do you know that Jews bless God for creating holes in their body? Do you know that higher order organisms have a greater variety of holes than lower order organism?
Dr. N: ... and I suppose that you consider yourself among the higher order organisms. I've had enough of those philosophical discussions. Sometimes I wish you had become "non-existent" in order to save me from your scholarly arguments.

Bamba: But then my presence would be much more evident through my non-presence! Think about it. Zero information may sometimes be extremely meaningful.

Chapter 12
Laws of Form

Summary: George Spencer-Brown developed a unique system of notation that uses distinction as its most basic unit. We may use this system in order to explain the way the world of appearances emerges out of the primordial experience. I advise the reader who is not mathematically oriented to skim through those few sections that may seem too purely mathematical.

In the previous chapter I described nothingness as the source of existence, and the old mystical tradition that suggests that something comes into being as a result of a differentiating activity. The venture of building a system that discloses this deep universal truth in its different manifestations underlies Spencer-Brown's seminal text, "Laws of Form."

In the preface to the 1979 edition of "Laws of Form," Spencer-Brown clearly states that his book is "a text book in mathematics, not in logic or philosophy." Indeed, the book may seem to be non-relevant to our epistemological quest because it deals with Boolean algebra and other issues that may seem to be of minor interest to the non-mathematician. However, this book is familiar to the systems research community, not only because it presents a non-numerical arithmetic (calculus of indication), but also because it develops a unique notational system that takes nothingness as its starting point and may serve in order to inquire into the logic of distinction that underlies the world of phenomena. Although, Spencer-Brown describes his book as a mathematical treatise, in the preface to the latest edition of the book (1994), it is described in mystical terms and as having wider relevance than just to the mathematical community. The reader doesn't have to wait for the 1994 edition in order to sense Spencer-Brown's desire to say something which is beyond the mathematical realm. At the AUM conference, held on 1973, Spencer-Brown was asked to explain what his book is about, and, when doing so, the discussion digressed into other fields far from the purely mathematical origins of "Laws of Form." One of his statements is especially interesting. In response to a question he says:

> In reality, it is all the same. In reality, it is a matter of indifference, but we are not here in reality. We are here on a system of assumptions, and we are all busy maintaining them. On that system, then we can say, "Well, that will keep the ship afloat, and this will pull the plug out and we will all sink." In reality, there is nothingness. An undifferentiated reality. However, when we differentiate reality, value judgement cannot be avoided in order to support systems of differentiation with their continuity.

In other words, distinctions should not be confused with reality. The world we experience as human beings is only one possible way of approaching reality among the infinite other possible systems of distinction. Other organisms see the world in a different way and we may say that they have a different *Umwelt* (von Uexkull, 2001). The world of appearances (= distinctions) and the signs (= indications) we attribute to these appearances (what will be later called second-order distinctions) is a world of one possible expansion of reality.

With regards to our reified universe, we should recall that *identity*, the identity of the sign with itself, is just one of those value judgements that allows us to keep on speaking about the things as if they were static and real objects having "real" existence beyond our minds, and not the invention of our differentiation system. This is an interesting perspective, which will occupy us further in the book. In short, we may argue that the identity of the sign (a pipe is a pipe) and its correspondence to a preceding object is a fantasy, and maybe the most important fantasy, that keeps our semiotic ship afloat.

At the beginning of his book, Spencer-Brown makes an important comment that links him to the mystical tradition previously discussed. He writes: "The theme of this book is that a universe comes into being when a space is severed or taken apart." More specifically:

> A distinction (something) is drawn by arranging a boundary with separate sides that a point on one side cannot reach the other side without crossing the boundary...Once a distinction is drawn, the spaces, states, or contents on each side of the boundary, being distinct, can be indicated. (p. 1)

Therefore, Spencer-Brown presents the *primary distinction*, symbolized in the current text by the mark "()" as its starting point and the non-marked state with an empty space.

Robertson (1999) nicely illustrates the abstract idea of the first distinction by asking the reader to imagine nothingness as a great flat sheet without any thickness, boundary or any other differentiating characteristics, which extends forever. The primary (or the first) distinction is illustrated as a circle (without any thickness) that differentiates this void. Now let us draw a circle that has no thickness. This circle may look approximately like that:

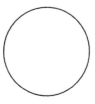

This circle has an inside, an outside, and a boundary that differentiates those units. The differentiation process is symbolized as an arrow from the inside to the outside:

Axioms are arbitrarily assumed. We do not have to justify them or to ground them in the world of appearances. Surprisingly, the two axioms presented by Spencer-Brown seem to contain a universal truth of sign systems. If we consider the primitive equations in the context of semiotics, we can immediately see their linkage to our semiotic experience. The form of condensation is associated with Saussure's idea of the sign as a differentiated unit. In a similar way to the distinction that has meaning only by being distinct, the sign has meaning by being distinct from the other signs in the system. Both the first distinction and the sign do not have intrinsic properties that define their identity. Their identity is determined by their distinctiveness.

Concerning the second axiom, a distinction (= difference) cannot be differentiated since it is the most basic unit of any analysis. The ontological status of the primary distinction explains the second axiom: A difference, which is differentiated, turns us back into the void. We may interpret the form of cancellation as saying that there is bipolar road from distinctiveness to nothingness. How can we interpret the form of cancellation in the context of signs? One possible interpretation is that the primordial difference is closed to further differentiation on the same ontological level. However, it can be differentiated by expanding *"upward"* or *"outside."* The dynamic of the primary distinction is such that it can be further differentiated by expanding from nothingness and not the other way around. Our world is differentiated, and can be differentiated more and more, but only from a certain ontological anchor which in itself cannot be differentiated.

3.1. FORM AND DYNAMIC

The forms described above lead to a calculus of expressions (primary arithmetic) in which expressions like:

$$(()())(())$$

may be reduced by cancellation and/or condensation to the marked or the unmarked state. Thus, the above expression may be reduced as follows:

Step 1. Through cancellation: (()())
Step 2. Through condensation: (())
Step 3. Through cancellation: to the unmarked state, which is the value of the above expression.

Kauffman and Varela (1980) suggest that we can conceive this process as if the *deepest space of an expression is sending signals of value* (marked/unmarked) through the expression to be combined into global valuation. If we denote the marked space by "m" and the unmarked state by "n," then the above expression can be described as:

Laws of Form

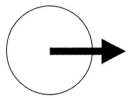

There is no explicit reason for choosing a circle [over a] triangle or other form. However, it seems that the attractive[ness of] the circle is a crucial factor in this aesthetic choice. The[re is a] possible explanation for choosing a circle, and this explanat[ion is] the somatic nature of our mind. In an essay, entitled "Ci[rcles,"] Waldo Emerson (1940, pp. 279-291) says: "The eye is the fi[rst;] horizon which it forms is the second; and throughout this prin[ciple is] repeated without end" (p. 279). Since our visual encounter w[ith reality] is through a circle, it is not surprising that a circle is used [for the] primary distinction! This circle is something that eme[rges from] nothingness by arranging a boundary. Through the si[mple act of] *differentiation*, something comes into being out of nothingne[ss. In this] analysis, we may infer that the most basic unit of existence in [an act] of differentiation, an idea that appears in the Book of Genesis [and] other ancient texts. I will argue later that this is not a single a[ct,] but a *continuous* process that constitutes the primary distincti[on. This] section aims to present Spencer-Brown's calculus of indic[ations and its] elaboration by Varela and Kauffman. The reader who feels u[neasy] with the mathematical spirit in these sections may skip [the] details while keeping track of the general arguments.

After presenting the act that brings something [forth,] Spencer-Brown presents two axioms: The law of callin[g (form of] condensation) and the law of crossing (form of cancellation). [The law of] condensation states that, "The value of a call made again is [the value of] the call" (p. 1), visually described as:

$$() \, () = ()$$

And the form of cancellation states that a distinct [re-entry of a] distinction brings us to the unmarked state, visually described [as:]

$$(()) = .$$

The first form states that two things are identical if [they are not] distinct, and the second form states that a mark can be a[nnulled by] turning on itself through the act of differentiation, in this spe[cific case] by making a cross within a cross. These basic forms are [called by] Spencer-Brown as the "primitive equations."

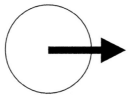

There is no explicit reason for choosing a circle rather than a triangle or other form. However, it seems that the attractive symmetry of the circle is a crucial factor in this aesthetic choice. There is another possible explanation for choosing a circle, and this explanation points at the somatic nature of our mind. In an essay, entitled "Circles," Ralph Waldo Emerson (1940, pp. 279-291) says: "The eye is the first circle; the horizon which it forms is the second; and throughout this primary figure is repeated without end" (p. 279). Since our visual encounter with the world is through a circle, it is not surprising that a circle is used to denote the primary distinction! This circle is something that emerges out of nothingness by arranging a boundary. Through the simple act of *differentiation*, something comes into being out of nothingness. From this analysis, we may infer that the most basic unit of existence involves an act of differentiation, an idea that appears in the Book of Genesis and in many other ancient texts. I will argue later that this is not a single act of creation but a *continuous* process that constitutes the primary distinction. The next section aims to present Spencer-Brown's calculus of indication and its elaboration by Varela and Kauffman. The reader who feels uncomfortable with the mathematical spirit in these sections may skip the technical details while keeping track of the general arguments.

After presenting the act that brings something into being, Spencer-Brown presents two axioms: The law of calling (form of condensation) and the law of crossing (form of cancellation). The form of condensation states that, "The value of a call made again is the value of the call" (p. 1), visually described as:

$$() \, () = ()$$

And the form of cancellation states that a distinction within a distinction brings us to the unmarked state, visually described as:

$$(()) = .$$

The first form states that two things are identical if they are not distinct, and the second form states that a mark can be annihilated by turning on itself through the act of differentiation, in this specific example by making a cross within a cross. These basic forms are described by Spencer-Brown as the "primitive equations."

Axioms are arbitrarily assumed. We do not have to justify them or to ground them in the world of appearances. Surprisingly, the two axioms presented by Spencer-Brown seem to contain a universal truth of sign systems. If we consider the primitive equations in the context of semiotics, we can immediately see their linkage to our semiotic experience. The form of condensation is associated with Saussure's idea of the sign as a differentiated unit. In a similar way to the distinction that has meaning only by being distinct, the sign has meaning by being distinct from the other signs in the system. Both the first distinction and the sign do not have intrinsic properties that define their identity. Their identity is determined by their distinctiveness.

Concerning the second axiom, a distinction (= difference) cannot be differentiated since it is the most basic unit of any analysis. The ontological status of the primary distinction explains the second axiom: A difference, which is differentiated, turns us back into the void. We may interpret the form of cancellation as saying that there is bipolar road from distinctiveness to nothingness. How can we interpret the form of cancellation in the context of signs? One possible interpretation is that the primordial difference is closed to further differentiation on the same ontological level. However, it can be differentiated by expanding *"upward"* or *"outside."* The dynamic of the primary distinction is such that it can be further differentiated by expanding from nothingness and not the other way around. Our world is differentiated, and can be differentiated more and more, but only from a certain ontological anchor which in itself cannot be differentiated.

3.1. FORM AND DYNAMIC

The forms described above lead to a calculus of expressions (primary arithmetic) in which expressions like:

$$(()())(())$$

may be reduced by cancellation and/or condensation to the marked or the unmarked state. Thus, the above expression may be reduced as follows:

Step 1. Through cancellation: (()())
Step 2. Through condensation: (())
Step 3. Through cancellation: to the unmarked state, which is the value of the above expression.

Kauffman and Varela (1980) suggest that we can conceive this process as if the *deepest space of an expression is sending signals of value* (marked/unmarked) through the expression to be combined into global valuation. If we denote the marked space by "m" and the unmarked state by "n," then the above expression can be described as:

$$(()m()m)n(()m)n$$

From this the value "n" emerges.

The calculus of indications can be extended to an algebra that deals with variables regardless of their value. For example, we can show that the form ((p)) is always equal to p: If p is the marked state then ((())) = (), and if p is the unmarked state then (()) = .

In this context, Spencer-Brown introduces the *imaginary* state f = (f) in which a form enters itself the same as the imaginary number – plus/minus I – emerges from the equation $x^2 + 1 = 0$. If f is () then f = (()) equals the unmarked state which gets at the next step into the equation and leads to f = (). This self-referential dynamic is highly important, first, since it was "prohibited" by Russell's famous Theory of Types, and Spencer-Brown is presenting a totally different perspective that legitimizes the use of self-referential terms in a system of logic. Second, it is important because self-reference introduces *memory* into the system. By stating that the value of a system is determined through self-reference, we must recall the last value of the system. That is, if f equals on T1 to "()" then on T2 it is equal to the unmarked state. It is as if the system remembers in what state it was a minute ago and uses this memory to determine the next state. What is so fascinating about the whole thing is that *memory is introduced to the system as a modus of self-referential activity and not as an exogenous factor*. Later I will argue that the boundary of the primary distinction, the boundary that defines each living system, involves this kind of self-referential activity. Therefore, memory is encapsulated in the notion of self-reference and in the intermediate meeting of the organism with its environment. The implication of this idea is that the world is not represented in our cognitive system through some simple form of mapping from the external world into the mind, but that the external world is evident in our systemic closure through memory, through its boundary relations with its environment.[36] We do not remember things that are a-priori distinct, but the world of appearances exists to us as a memory which is a modus of a differentiation activity, of our relationship with the environment. This conception is totally different from the conceptions of memory evident in Plato's conception of memory, the conception of memory in classic cognitive science, or even the highly fashioned connectionist conception of memory (Varela, 1992). In Spencer-Brown's system of thought *memory is a form of being, of a distinction, of a self-referential activity, of a boundary*.

The self-referential activity also introduces the most primitive notion of *time* we have. This is a primordial time. As suggested at the AUM conference by Spencer-Brown, the intervals of this primordial time are "neither short nor long." It has no duration at all, no determined frequency. It has no measure. Similarly to the primary distinction, which is not spatial, primary time is not segmented. Both establish their existence by Being, in the purest sense of the term.[37]

Inquiring further into Brown's system, Kauffman and Varela call our attention to the relation between *form* and *dynamic* that is evident in the calculus of indications:

The geometric form of the expression represents the figures; calculational steps are an elementary dynamics; signals of values moving through the expression represents a kind of periodic vibration. (p. 7)

They show that when we allow an expression to contain a variable that changes its value in time (i.e., vibrates), some strange and fascinating things can happen. From our point of view, this is an important theoretical move since I would like to argue that the boundary that defines the primary distinction is a vibrating and indeterminate activity. I will follow the example used by Kauffman and Varela in exposing the consequences of the vibrating variable.

Consider the expression: (((a))). If a = n, that is, if our variable equals the unmarked state, then the whole expression e is:

$$(((n)m)n)m$$

If a = m, then the expression is:

$$(((m)n)m)n$$

So, at a certain point we would see something like:

$$(), (()), ((())), (((()))), ((((())))), \ldots$$

That is, an expanding form in which the deepest space is indeterminate due to its vibrating nature. That means that looking inside the primary distinction, one may see only the ultimate whole, the one, nothingness. Looking outside the primary distinction one may see the expansion of the distinctions into enormous divergences of distinctions and multiplicity. A divergence of appearances. The interesting implication of the above analysis is that the infinite temporal form f re-enters itself:

$$((((((v)))))) = f$$

Where this self reference may be described as a prescription for a recursive action

$$f \rightarrow (f)$$

and thus

Laws of Form

$$() \rightarrow (()) \rightarrow ((()))$$

therefore, the

> vibration of the deepest space yields self-referential spatial form, while the associated recursive dynamics to the self-reference unfolds the vibration (once gain) into a temporal oscillation. (p. 9)

The above statement is highly important since it suggests that the vibration of deepest space, its indeterminacy, results in a self-referential form. As will later be described, *this is exactly the boundary that constitutes the primary distinction between being and non-being*.

Apparently, if we are dealing with the infinite form f, Kauffman and Varela's analysis presents a paradox:

$$f = (f) = (ff) = ((f)f)$$

Hence:

$$f = (f) = f = .$$

Which means that f = ., i.e., the marked form f equals the unmarked state. Kauffman and Varela's solution to the above paradox is that Brown's calculus of indication considers the marked state as purely spatial without any temporal meaning. However, this paradox may be solved if we consider the first distinction as having a temporal sense. Varela himself (1979) suggests adopting a third state (the autonomous state or self-cross), distinct from the marked and the unmarked state. By introducing the third form, Varela moves from a two-value logic (the marked and the unmarked state) to a three-value logic.

Intellectually intriguing as it is, the calculus of indication and its extension by Kauffman, Varela and others, leaves open some of the main problems facing us when we try to understand the self-referential nature of the mind. The mind is not infinite, but finite, and should be examined according to its unique ability to unfold *meaning* and not to elaborate the truth-value of a given proposition. In this context, introducing the idea of self-reference, as an integral part of the theory does not clarify the nature of the mind. By introducing mystery into a theory one does not explicate the mystery, but just legitimizes it. Legitimization is an important act. However, the mystery of self-reference exists whether we legitimize it or not, whether we introduce it as an irreducible idea to our theory or not, and whether we formalize it or not.

The question is whether by introducing the ideas of self-reference and the primary distinction into our analysis we can better understand the phenomenon under inquiry. My answer is definitely positive. Spencer-Brown "Laws of Form" provides us with insightful ideas how to conduct our inquiry. However, there are several difficulties associated with the calculus of indication and its later developments.

The most worrying thing about Spencer-Brown's first distinction is that the status of the *observer* in it is not clear (see also Glanville, 1990). Indeed, there is someone (or something) who makes a distinction. The construction of the first distinction is even manifested by Spencer-Brown as a command for a given observer: "Draw a distinction." But who is the one who draws the distinction? If there is someone or something that makes the distinction, then the first distinction is not really the "*first*" distinction. It is either a first distinction made within an abstract and artificial system of notation or a first distinction made by a divine force beyond the grasp of our intellectual ability. In both cases, nothing scientific can be said about the mind. The answer to this difficulty is that *each living being is a primary distinction*. Each of us and all the other living systems around us are primary distinctions that operate according to the same logic of distinction. We will get to this point later after elaborating further the logic of distinction.

Cat-logue 6
Inside the Outside

Dr. N: I would like to share with you a nice piece I read lately. In one of his "Knots" R.D. Laing writes something that should be remembered while considering a living system as a primary distinction (1970, p. 83):

> One is inside
> then outside what one has been inside
> One feels empty
> because there is nothing inside oneself
> One tries to get inside oneself
>   that inside of the outside
>   that one was once inside
>   once one tries to get oneself inside what
>   one is outside;
>   to eat and to be eaten
> to have the outside inside and to be
>   inside the outside.

Bamba: My goodness, this shrink knows how to play with words, but what the hell does he mean?
Dr. N: You're asking me? Well, if we consider ourselves in simple spatial terms, then I am inside and my environment is outside.
Bamba: Clear as crystal. Don't tell me that you are going to argue with this truth?
Dr. N: This is exactly my aim. I argue that we are (or more accurately our mind) neither inside nor outside, but a boundary phenomenon. We are always in between.
Bamba: This statement is so obscure that I assume that you inherited some obscure genes from the late Prof. Laing.
Dr. N: No, but I am a relative of John von Neumann and I assume that I share some of his "scientific" genes.
Bamba: In this case, I expect you to be more scientific than you are.
Dr. N: But what I am saying is pure science. Just read Laing's knot. If we are inside then the only way of seeing ourselves, or reflecting on ourselves is from the outside and we must see ourselves from the outside in order to differentiate ourselves from the outside. This is how the primary distinction is associated with semiotics. Since without signifying the outside there is no inside and vice versa.

Bakhtin (1990) wrote about this in one of his earlier philosophical texts. Just read:

> A very special case of seeing my exterior is looking at myself in a mirror. It would appear that in this case we see ourselves directly. But this is not so. We remain within ourselves and we see only our own reflection, which is not able of becoming an immediate moment in our seeing and experiencing of the world. We see the *reflection* of our exterior, but not *ourselves* in terms of our exterior... I am in front of the mirror and not in it. The mirror can do no more than provide the material for self-objectification, and even that not in its pure form. (p. 32)

Bamba: What does he mean?

Dr. N: He means the same thing as Laing: We cannot grasp ourselves, or more accurately, our exterior, from within, only from the outside. This is usually done through a variety of "mirroring devices" that tell you who you are. These mirroring or representation devices are for example *other human beings*. Remember this idea, because this is the source of our reified universe. While a sign system generates signs out of distinctions, the others cause us to fix those signs and provide us with the ideology that justifies this process of objectification. The problem is that those devices do not let you see who you are, rather only your reflection.

Bamba: Now I see why Laing suggests that looking for ourselves outside is problematic because there is nothing outside.

Dr. N: Correct, and then one tries to get inside ...

Bamba: ... from the outside?

Dr. N: Yes, and then he finds nothing.

Bamba: Because the inside is ...

Dr. N: Inside, which you cannot see from the inside, but only from the outside, which brings you again to the mysterious oscillation we just portrayed. You see, the flux of being we portrayed in the first part of the book cannot be grasped from outside by using re-presentations that by definition cannot grasp the dynamic of live experience.

Bamba: Now I see the way the reified universe is constructed! By looking at us from the outside and through different forms of semiotic mediations, that fix this awful oscillation between the inside and the outside!

Dr. N: Correct. To recall Bakhtin again: "The mirror can do no more than provide the material for self-objectification." You see, "*self-objectification.*"

Bamba: This is a terrible conclusion since it destroys the whole existentialist movement in psychotherapy that tries to look for the authentic self inside us.

Dr. N: I have been looking for this authentic self since I was a teenager and I never did find it inside. I don't even know where this "inside" exists.

Bamba: Try to look for the inside outside.

Dr. N: Clever. Are you some kind of a Rogerian therapist or just ELIZA the artificial therapist?

Bamba: I am BAMBA the catherapist. Would you like to talk about it?

Chapter 13
Toward a Phenomenology of Boundaries

Summary: The boundary is a socio-somatic-semiotic dynamics that constitutes the observer and her environment. The boundary is self-referential, indeterminate and vibrating activity.

Let us go back to the visual illustration of the primary distinction:

Herbst (1993) suggests that the primary distinction is a unit consisting of no less than three elements: the inside, the outside and the boundary that may be denoted in the most general sense as [n, m, p]. The divine trinity is a recurrent theme in human culture: The Christian Trinity, Hegel's thesis, antithesis and synthesis, Peirce's Firstness, Secondness and Thirdness (e.g., icon, index, symbol), Freud's Id, Ego and Superego, and Merleau-Ponty's three orders of signification, matter, life and mind, are only a few examples of the mania for three described by Peirce as "Triadomany." As a monotheistic Jew, Herbst would have been delighted to realize that the triadic structure of the primary distinction can be reduced to a single principle: The vibrating nature of the boundary. However, at that time, Herbst had different ideas in his mind and it is worth inquiring into them before presenting ours.

The first property of the primary distinction is that it is "co-genetic" (Herbst, 1993) in the sense that the three elements – n, m, p – "come into being together" (p. 30). I prefer to use the term "co-generic" because it does not bear the flavor of genetic research and emphasizes the generic power of the primary distinction. Beyond words, the meaning of this property is that the basic unit of the mind is something that *emerges* out of a relational structure and not from either a material or other single and homogenous atom. To recall, the basic unit of the mind according to the Cartesian philosophy of Rene Descartes is some kind of immaterial self and not the synergetic product of sub-units. Indeed, there are radical conceptions of the self, such as that of Bakhtin, that argue exactly the same thing: The self is a synergetic product that emerges out of a relational structure: self and non-self. We will get to this idea later in our discussion. Meanwhile, we should just keep in mind Herbst's suggestion that the primary distinction is composed out of three units that do not have an independent existence but that come into being only together. The first distinction is an *integrated whole*.

The second property of the primary distinction is that it is unseparable in the sense that we cannot take the units apart. This is an important property because it states that wholes, as distinct macro-structures, are resistant to any form of simple reductionism. This property is supported by the third property that states that the primary distinction is non-reducible. It is a whole that cannot be reduced to its components. This idea clearly adheres to the idea of a *Gestalt* in psychology. However, while a Gestalt can be reduced to its components such as a circle that can be theoretically reduced to points, the primary distinction cannot be broken into any meaningful components. Its existence is holistic in the ultimate sense of the term.

The third and last property discussed by Herbst is that the primary distinction is contextual in the sense that the units composing the primary distinction have no intrinsic characteristics. They are not units with some kind of an Aristotelian essence that determines their identity. The identity of each unit is determined only by its relation to the other units: There is no inside without outside, there is no outside without inside, and there is no inside and outside without a boundary that differentiates between the two units and therefore constitutes their existence.

I would like to add to this list a fourth property, which is adopted from Bateson's idea of a recursive hierarchy. The property I would like to discuss is the *recursive-hierarchical* structure of the primary distinction. The primary distinction is an integrated whole, which is composed of three subordinate units. In this sense, it is a hierarchical structure with complex synergetic whole and subordinate, less complex units. However, the primary distinction is far from being a static hierarchical structure of a set and its components. The meaning of the whole structure is determined by the dynamic of its units. Only by operating together does the whole structure emerge. On the other hand, the meaning of each unit can be determined only by the whole of which it is a part. At this point, we encounter again the mythological serpent that eats its own tail, and arrives at what may look like a vicious circularity: The meaning of the units is determined by the whole and the meaning of the whole is determined by the units. Is it the same as trying to determine A by B and B by A? Not exactly. Indeed, the structure of the primary distinction is circular, but, due to its hierarchical structure, it involves two differentiated levels of analysis: the whole and its units. Therefore, we are not speaking about simple circularity that exists on a single level of analysis, but about a recursive-hierarchical structure in which the whole generates the meaning of its sub-units and the sub-units generate the meaning of the whole in a continuous process. This process constitutes the Holon we describe as the primary distinction. This process may be described as a recursive-hierarchy, but it was also known for many years as a *hermeneutic circle*. The term may be easily illustrated in the context of textual interpretation. Let us assume that we try to understand the meaning of a specific sub-unit of a text. For example, we read a text and try to understand the meaning of the sign "the cat." How do we determine what the specific meaning of this sign is? Is it an animal? The name the local press gave to a skillful burglar? The name of a new punk band?

To recall Volosinov, in itself the sign is devoid of meaning. In order to understand the meaning of this sign we can consult a dictionary. A useless attempt, as we all know. In order to understand the meaning of this sign we need to understand the whole text of which this sign is a part. Hey! But how can we understand the whole text without knowing the meaning of its components? The answer is that the process of understanding a text and a sign is a recursive hierarchical process in which the whole text provides the meaning of its sub-units (the signs) and in return the sub-units provide the meaning of the whole text. This self-referential structure is the basic structure of cognition and of all living forms. Trying to deal with it through first-order logic or through other "flat" tools of interpretation simply does not work. This process of understanding is known as "hermeneutic circle" and it is further elaborated in the next section.

13.1. THE HERMENEUTIC CIRCLE

Hermeneutics can generally be defined as the philosophy of interpretation. Remember Hermes? The wing-footed little guy from Greek mythology? Hermes delivered and interpreted messages from the gods to mortals, and the field of interpretation was named after him. However, why do we need a theory of interpretation in the first place? Is interpretation a human need, the same as food and sex? Trying to deal with needs has an essential peculiarity I would like to avoid in this book. However, it seems that the dynamic of the human sign system is such that each text must have a context in order to become meaningful. In other words, since human beings do not process information but somehow elaborate meaning, interpretation through text-context dynamics is an inevitable aspect of their existence. Historically we know that matters of interpretation become especially acute when we try to deal with translation from language to language, by trying to restore a dead language, when authoritative figures who "interpret" the text are challenged, and when old methods of understanding a text seem to crack. Interpretation as a conscious activity is associated with a break in our ability to participate in meaning-making practices. Historically, we also know that hermeneutics enlarged its scope from local philological conceptions of biblical interpretation to Schleiermacher's conception of hermeneutics as the foundation for all kinds of textual interpretation and Dilthey's ideal of hermeneutics as the general foundation of human sciences. Schleiermacher was the one who formally introduced the idea that the meaning of the part is always discovered by the context and that understanding is necessarily circular. A similar notion about the hermeneutic circle has been presented and elaborated by Gregory Bateson. In his conclusive analysis of Bateson's concept of recursive-hierarchy, Harries-Jones (1995) calls our attention to the fact that "in order for any information to have meaning, the information messages must be framed and contextualized by several different interacting and integrated levels of the senses – speech, sight, hearing, "body language" – *and by patterns of interaction among members of the communicating unit* (p. 245, my emphasis) This point has been elaborated in the previous sections, but it is worth mentioning it while discussing the primary distinction: Meaning is created when a difference (always attached to our somatic-fleshy base) is contextualized by the interacting agents. Another important feature of informational order is that "for any level to remain in stable relations with any other level, it cannot 'stay still.'...In short, a communicative order must maintain a *meta-stability* in relation to a supporting environment through constant variations of relations between its levels" (p. 245). That is, the nature of every living system is organized in a multileveled and dynamic structure as evident from the structure of the primary distinction. Following this line of thought we may proceed and further inquire into the structure of the primary distinction.

The boundary plays a crucial role in defining a systemic closure such as the primary distinction. But what is a boundary? Does it belong to the differentiated form or to the environment? For several reasons, the topological concept of a boundary as used by mathematicians (especially in point-set theory) is not well suited for describing boundaries in living systems. Mathematical topology is closely associated with set theory and uses points as fundamentals. The difficulties of applying set theory to the study of communication in living systems are well documented. In addition, due to their highly abstract mathematical nature, points do not seem to be the relevant fundamentals for the analysis of living systems that are composed of cells, people or signs. Therefore, the topological concept of a boundary is of little relevance to our task. A possible and more relevant interpretation of the boundary is as an oscillatory process. The next sections present the idea of the boundary as a process that constitutes the existence of the primary distinction.

13.2. BOUNDARIES

If we examine the boundary closely from a phenomenological perspective, we can see that its *value* (whether it belongs to the differentiated system or to the environment) cannot be determined. The boundary exists as long as we oscillate between the system and its environment. The boundary is neither the observer nor the environment. It is a vibrating activity that constitutes, through its indeterminate moments, the observer and its environment, the figure and the ground, the marked and the unmarked states. Janus-faced, it looks toward being and non-being alike; it is the vertex of being and non-being. In cybernetic terms, the boundary may be described as an *oscillation,* as a periodic change of the values (e.g., figure and ground, self and non-self) of a given function. Therefore, from a phenomenological perspective the boundary is not parasitic of the entities it defines, but precedes and constitutes them through its indeterminacy.

It must be noted that the boundary has a paradoxical/self-referential structure, since its value always re-enters itself: if it is inside, then it is outside and so on. We must shift continuously between the two differentiated units (e.g., the figure and the background) in order to grasp each entity in itself. As I previously said, it was Spencer-Brown who insightfully argued at the AUM conference[38] that this simple self-referential expression introduces the concept of *memory* into the system, since the state of the system at step n+1 is determined by its state at step n. In this sense, boundary as an oscillatory function is not only a differentiating process (a *dynamic*) that constitutes differentiated *forms* (e.g., the being and his or her environment) but also a mode for transferring the "outside" into the "inside" (and vice versa), though not in a simplistic, representational manner, as usually portrayed in orthodox cognitive science. The environment is evident in the observer (the system we describe as the loci of the cognitive activity) not through a mapping of the external world onto the mind but through memorization of the environment, which is the boundary's mode of periodic movement. This idea suggests that memory is not the simple recording of external/internal events but a mode of the boundary. A signification process that constitutes the observer and the environment. *Following this line of thought, the study of meaning-making in living systems is primarily a phenomenology of boundaries within a recursive-hierarchical structure.*

The question that emerges from the foregoing analysis is how a self-referential, oscillatory process (boundary) can explain the emergence of meaning (the structured world of appearances) in a recursive-hierarchical structure. To answer this question, we should introduce the idea of the body as an ensemble of sheets or surfaces.

In a recent manuscript, Engelberg (2001) argues: "Multicellular organisms, at virtually all stages of their life cycles, are ensembles of two-dimensional structures (sheets) of various kinds" (p. 1). Engelberg defines a sheet as an object whose length and breadth are much greater than its thickness. According to this suggestion, a systemic closure, like an onion, does not necessarily involve a single primary boundary that clearly defines its closure, but an ensemble of multi-layered surfaces that together constitute the differentiated organism. If we apply this to our case, the observer may be considered a multi-layered (hierarchical) structure of boundaries. In this sense, there is no point in looking for the "self" beyond the boundaries that define the systemic closure.

The idea that living systems operate according to a multi-level boundary structure is a first step in understanding the emergence of meaning in those systems. Maturana et al. (1995) have suggested that a recursive process is evident when a linear process encounters a circular process:

> When a repeating circular process becomes coupled with a linear one that displaces the circumstances of the repetition, the repetition of the circular process becomes a recursion and a new phenomenal dimension appears. Thus, for example, when the circular movement of the wheels of a car is coupled with the linear displacement of the ground, the circular movement of the wheels becomes recursive and the phenomenon of movement appears. (p. 3)

This is an interesting suggestion, but the meaning of linearity, circularity or the process through which a recursive process becomes evident is not well elaborated. It may be theoretically useful to replace the idea of a linear process with the idea of *periodicity,* since linear processes are rare and "Events ... do not take place in a continuous sequence, in a straight line, but are in a continual state of vibration, oscillation, undulation and pulsation" (Jenny, 1974, p. 10).

The external environment continuously "sends"[39] the organism pulses of energy in a periodic form. The same is true of human communication systems in which the subject is continuously exposed to repeated patterns of interaction (periods). The coupling of these periodic processes with the unique multi-surface (boundary) structure of the organism turns this information/periodicity into "meaning," or what Merleau-Ponty calls "habit." When our boundary structures interact with the environment, this interaction echoes through the oscillatory and hierarchical nature of the organism into a higher/deeper surface. Thus, we can see that the whole of the organism as a differentiated systemic closure may be traced to its multi-layered and recursive dynamic.

In short, the basic unit of the mind is a Holon we describe as the primary distinction. This Holon is constituted through a vibrating boundary (actually boundaries) that defines the meaning of the systemic closure and its environment. This Holon is a recursive hierarchical structure in which meaning is produced in-between the levels. If we are such Holons, then certain questions concerning our mind as a semiotic interface may be clarified. One of those ideas is the idea that a sign is a distinction and, therefore, also a boundary phenomenon constituted by the same laws of the primary distinction. If we consider ourselves through a semiotic perspective then we are, as suggested by Peirce, signs. If I am a sign, and a sign is a boundary that exists in-between, what does it mean concerning my identity? Where do "I" exist if there is no real me inside of me?

Cat-logue 7
How Deep is the Surface

Bamba: Your presentation of the sign as a boundary phenomena really bothers me.
Dr. N: Really! Why?
Bamba: First and foremost because it contains an implication for a theory of meaning, and meaning is a concept that has bothered us from the first pages of this book.
Dr. N: Well, you are right. The meaning of a sign bothers us since it is the key for understanding our reified universe. If the meaning of the sign is not referential, we may question the existence of the objects at which it points. However, don't you think that it would be helpful to clear out the meaning of meaning before discussing the meaning of the sign as a boundary phenomenon? You must be familiar with the long scholarly tradition, originating in classical Greek and epitomized in the writings of Aristotle, that required that a scholar completely conclude any discussion and conceptual clarifications before presenting his own thesis of a given issue.
Bamba: Well ... I am well aware of this tradition. However, this long tradition was relevant to a period in which the amount of information concerning a given issue was quite limited. With your permission, I prefer to get to the point.
Dr. N: Ok, my favorite cat, what bothers you so much?
Bamba: Human beings and the Western culture I am familiar with attribute more importance to depth than to surface. All the metaphors we use, "deep thought," "deep structure," etc., attribute positive value to depth and negative value to surface.
Dr. N: Correct. Wisdom is at the top of the mountain (e.g., Moses and Zarathustra) or at the bottom of the sea. Don't you think that it has something to do with the Platonic ideal of appearance = surface and reality = depth?
Bamba: I was just thinking about this linkage. Plato's vision of science clearly adheres to the depth/surface metaphor. However, your exposition of the boundary as an activity constituting the world of appearance *attributes much more importance to the surface than to the depth!*
Dr. N: Well ... Hmm ... This is a provocative thesis, but I don't see why. From where did you adopt this idea?
Bamba: From the depth of my soul! I'm joking! Your spatial-metaphorical language is too simplistic. When you ask "from where," do you want me to name the "place" in which this idea originated? The location in my brain? The reference I quote? I thought that we agreed to avoid simple metaphors.

Dr. N: As you know Nietzsche has already said that we are prisoners of our metaphors. So, please describe to me the "event" in which this interesting idea popped into your mind.

Bamba: I was reading "The Logic of Sense" (1990) by Deleuze and ...

Dr. N: My God!!! After blaming me for being a Post-Modernist *you* were reading a book by a French Post-Modernist! Did you read Sokal's book, "Fashionable Nonsense," before approaching a Post-Modernist text? I am sure that you did not understand a word.

Bamba: At this point of my intellectual development, I must admit that I found the book very interesting.

Dr. N: Indeed, it is a brilliant intellectual achievement.

Bamba: Yes. I know that you agree with me since in this book I am playing the role of your alter ego. But let's go back to the issue. Deleuze is saying that "Paradox appears as a dismissal of depth, a display of events at the surface, and a deployment of language along this limit" (p. 9).

Dr. N: Now I see the link. Deleuze is also pointing at the fact that "Good sense affirms that in all things there is a determinate sense or direction, but paradox is the affirmation of both senses or directions at the same time" (p. 1). Since the boundary, as I described it before, involves a paradoxical movement through its oscillation, it is a surface phenomenon.

Bamba: Correct, and what is implied from this analysis is that meaning is an event, a paradox, a surface phenomenon. In the next chapter, you are going to examine the relevance of this idea to the meaning of the "I."

Dr. N: Astonishing, Holmes! How do you know what am I going to describe next?

Bamba: Well, Dr. Watson ... your surprise assumes the linearity of text and thought, which is not the case. In our dialogical imagination, your image of meaning precedes our dialogue.

Chapter 14
Peter Pan's Shadow and the Empty Observer

Much as Peter Pan's shadow is sewn to his body, the "I" is the needle that stitches the abstraction of language [the sign "I"] to the particularity of the lived experience.
Holquist

Summary: The idea that the observer, as a systemic closure, is identified with the primary distinction has some radical implications for the concept of "self" or "I." Those implications are discussed in the current chapter and aim to dismiss the Solipsist Cartesian concept of the "I."

In Chapter 3, I described the difficulty associated with Spencer-Brown's conception of the primary distinction. The solution to the perplexity described in this chapter is intelligible if we conceive the observer *as a modus of a differentiation process.* In other words, any system we may describe as an "observer" is a system that comes into being whenever it differentiates itself from the environment. The first distinction is between someone (or something) we may describe as an *observer* and her *environment*. However, stones, chairs and umbrellas are also differentiated entities. Do they deserve the term "observers," or "cognitive systems"? The answer is definitely no. Stones, chairs and umbrellas exist in our world as long as they are passively differentiated by external or internal forces and conceived as differentiated by observers. In contrast, observers operate according to a *boundary structure* that actively constitutes their identity as differentiated from their environment.

A skeptic may ask us how we know that an umbrella cannot differentiate itself from the environment. For this sophist question there is an answer. At the most basic level, an observer – a cognitive system – a living creature, may be defined as a closure that *actively* constructs its identity by behaving or reacting to physical forces as if they were a *difference that makes a difference,* a system that converts *information* into a phenomenal experience (= meaning). This idea brings us to Gregory Bateson and to the idea of a difference that makes a difference.

Bateson (2000) argues that while the natural world is a world of substance:

> when you enter the world of communication, organization, etc. [that is the world of living systems], you leave behind the whole world in which effects are brought about by forces and impacts and energy exchange. You enter a world in which "effects" - and I am not sure one should still use the same word - are brought about by difference. (p. 458)

Although there is a variety of differences in our environment (inside or outside) *the "elementary unit of information – is a difference that makes a difference"* (p. 459): My body temperature may fluctuates continuously, but only when it deviates from certain limits embodied in the biological system does this physiological difference (= information) becomes news (= meaning) and cause my behavior to change. Only then is it a difference that makes a difference. In this sense, and to use a phenomenological term, the environment lies before us as an "indeterminate horizon." Only by paying attention to certain features of it do we turn to a meaning-making practice, or the *making of our phenomenal world*. This activity may be better clarified by using the concept of "attention," a concept that has been described by cognitive psychologists as the spotlight of cognition. If we examine attention in the context of classic empiricist or rationalist paradigm, we get ourselves into deep trouble. To recall Merleau-Ponty again: "empiricism cannot see that we need to know what we are looking for, otherwise we would not be looking for it, and intellectualism [rationalism] fails to see that we need to be ignorant of what we are looking for, or equally again we should not be searching" (p. 28). Merleau-Ponty's solution to this perplexity is that attention is the "active constitution of a new object which makes explicit and articulate what was until then presented as no more than an *indeterminate horizon*" (p. 30, my emphasis). The world comes into being only at the boundary, at the *intersection* (i.e., the surface) of the subject and the dynamic object. Locating this statement in the context of the primary distinction we may say that the world comes into being only as a differentiation process that results from the dynamic of distinctions.

Out of the above phenomenological interpretation of a difference that makes a difference, we can clearly see that *meaning* (in the sense of a structured state of mind) is a difference that makes a difference, which is evident through change in the behavior of the system under inquiry. A stone is the object of physical forces in its environment. Wind, sun and flowing waters clearly change the structure of the stone. However, the stone does not change its "behavior" as a reaction to those forces. In a poetic sense, we may say that the stone is indifferent (in-different!) to its environment. In contrast, living systems are characterized by their behavior, a change of appearance that clearly corresponds to the environment. In a situation where a strong wind blows in your face, you look for shelter. In a situation where the sun burns your skin, you may move into the shade. Flowing waters may cause you to swim to the bank of the river or to lose your life, which is a behavioral change in its ultimate sense.

A boundary, an oscillation process, a difference that makes a difference stands at the heart of our phenomenal world. The radical implication of this idea is that our most basic interaction with reality precedes the existence of objects that obey the law of identity. The non-existence of identity (and therefore objects) at the most basic forms of being is supported by a beautiful and insightful experiment conducted by Köhler (described in Luria, 1976). The experiment was not designed in order to test the issue of identity, but I believe that it perfectly fits our case. A hen was presented with grains on two sheets of paper, one light gray and the other dark gray. On the light gray sheet, the grains simply rested on the surface of the paper, so that the chicken could peck at them, whereas those on the dark gray sheet of paper were glued in place so that the chicken could not peck at them. After being exposed to the sheets at several trials, the chicken learned the trick. It pecked at the light gray sheet and avoided the dark gray sheet. At this phase, Köhler turned to the crucial experiment and presented the hen with a new pair of sheets, one of which was the same light gray sheet and the other a new white sheet. Now the interesting question was how would the chicken behave in this case and to which of the sheets would she positively react? One possibility is that our hen is well familiar with logic. It is wired according to the law of identity. And it lives in a reified universe of objects identified with themselves. As my logic professor once said to us: Without accepting the law of identity nothing can be said or thought. If our chicken would have attended my professor's class, it should have positively responded to the light gray sheet. After all, this sheet is identical with itself and it is the sheet that brought her concrete benefits.

The results of Köhler's study are fascinating. In most of the cases, our chicken approached the *new white sheet*. Farewell to the law of identity! As some of us have assumed, chickens do not operate according to the abstract law of identity and in this case for good reasons. Köhler's explanation is that the hen had been directed not to the absolute darkness or lightness but to the *relative* darkness as implied by the *whole* situation and not by any particular component. This is the same conclusion I suggested previously concerning the relational structure of the mind and the cognitive precedence of the act over the object. Köhler's experiment is relevant for understanding our mind, since when we encounter the world at its most basic level, we operate according to the same logic.

The idea of a difference that makes a difference, with all its appealing simplicity, is an important idea that expresses a universal characteristic of living systems. It is the first building block of our epistemology that shed light on Descartes' epistemological obscurities. It is not that I exist because I think. "I" exist because I differentiate. Kauffman and Varela (1980) expressed it nicely: "We become individuals by making distinctions; the distinctions we make reveal (and sometimes conceal) who we really are" (p. 5).

We will get rid out of the Cartesian "I" in a later phase of our inquiry. For the time being, we should just acknowledge the fact that Husserl's epistemological wet dream can be realized in the realm of the primary distinction, the realm of a difference that makes a difference. In order to illustrate the enormous potential of a difference that makes a difference for phenomenological research in general and for our specific agenda in particular, let us dwell a little bit more on this unique link.

When I attune myself to a portion of the world (to the indeterminate horizon of dynamic objects), it thereafter turns into a difference that makes a difference, I allow the object to display itself (to correspond to my presence) by synchronically turning that portion of the world toward which I am not attuned into a "background." However, this background/context/frame tells us something interesting about the object. Paraphrasing one of Bakhtin's statements, it tells us that the object is always unique, but never alone. It comes into being only by reflecting its relations with other objects. "The completed object is translucent, being shot through from all sides by an infinite number of present scrutinies which intersect in its depth leaving nothing hidden" (Merleau-Ponty, 1962, p. 69). The object thus necessarily becomes visible through the invisible (the frame) and thereby turns into the suppressor of the other objects that exist in my phenomenal field. This is the same as we previously argued concerning the sign. When we turn our attention to our sign system, we tend to suppress the frame from which this system emerged, to repress its origins in the primordial realm of nothingness and activity, and to turn it into a system of self-contained "objects." In this sense, the reification of the world is a modus of our semiotic form of being.

This dynamic is the same when I turn my relations with the world – the atoms of my being-in-the-world – into an object of reflection, thus superseding my phenomenal field into a "mere background" and glorifying the object, i.e., the sign, the vehicle that turns on its condition of being, as if it were secondary to the existence of the object. This is the point where rationalists, who glorify the idea over the world, and empiricists, who glorify the world over the idea, both fail to provide a convincing answer to the mystery of the sign. Both fail to transcend the dichotomy of mind and the world, and to adopt the idea of mind as semiotic becoming that emerges as the boundary between the organism and its environment. This specific point calls for a further inquiry into the relationship between the mind and the world. Does the difference that makes a difference exist "out" there in the world or "inside" my mind?

14.1. IS THERE A DIFFERENCE THAT MAKES A DIFFERENCE IN OUR HEAD?

The idea of a difference that makes a difference previously presented may bring us to the question of whether there is some kind of a small person in our skull – homunculus – for whom there is a difference that makes a difference. We may rephrase this question as follows: Is the difference in our head? Answering this question may shed some light on the mind-world relationship. The phenomenological position previously presented suggests that we cannot describe the difference that makes a difference as something that happens in our mind, but as *mind itself*! In other words, the difference makes a difference whenever there is a correspondence between one system (= the observer) and another system (= the environment). In this context, it must be stressed once more that the primary distinction (the observer) first acquires an *operational* meaning rather than a semantic (intentional or extensional) one. By that I mean that we cannot consider the meaning of the primary distinction in terms of classic set theory according to the list of features that characterizes the observer (as a unique set), or by describing some kind of a general property that characterizes the set of observers. Anyone who has tried to get a hold of "Man's nature" (or essence) by listing features that characterize men has clearly experienced this difficulty. When I was an undergraduate student in psychology and philosophy, I had a talk with another student, who at the same time was a doctoral student in the biology department. After discussing some psychological theses with him, I ended the discussion by mentioning the concept "Man's nature." This doctoral student looked at me with surprise and assured me that as a biologist he was not familiar with the term "nature" as concerns men and particularly not with the unique combination "Man's nature." This interchange made a strong impression on me which lasted for many years. If biologists know nothing about Man's nature, what is the meaning of Man's nature in other domains of inquiry?

From the above interpretation of the mind-world relationship we conclude that the mind is the correspondence between the subject and his world. In other words, "The mental world – the mind – is not limited by the skin" (Bateson, 1973, p. 460), but exists as the correspondence between an observer and observed systems. From the above analysis, we may also conclude that the Kantian *thing-in-itself* is not only beyond our grasp but a meaningless concept. It is the correspondence between two systems (observer and environment) that constitute the mind and therefore the "things" that populate our mind and not the other way around. Since the world comes into being only through the attention of the observer, and the observer comes into being only through her attention to the world, what does it mean to have a "self" or an "I"? The next section tries to deal with this problem through Bakhtin, Volosinov and Peirce.

14.2. I AM A DIFFERENCE

Western thought has been enormously influenced by the Cartesian concept that the most basic and unshakable source of existence is the immaterial "I": *Cogito ergo sum*. It is interesting to examine the differences between this conception and notions of being derived from the conceptions of the mind as a semiotic interface.

To recall, Volosinov developed the idea that psychic life should be studied as a semiotic and dialogical phenomenon. As he wrote in

"Marxism and the Philosophy of Language": "Outside the material of signs there is no psyche" (1986, p. 26). This idea, which is over-whelmingly similar to ideas presented in the writings of Peirce, may be applied to the concept of "I." According to Bakhtin, the "I" is a sign that has no referent in the sense other signs (such as "tree") have. It does not point at anything or, to be more concrete, it points at nothing. In this context, the status of the "I" as the ultimate source of existence should be clarified: If the sign "I" points at nothing, what is the meaning of having an "I" or a "self"? For Bakhtin, existence is the event of co-being: "It is a vast web of interconnections each and all of which are linked as participants in an event whose totality is so immense that no single one of us can ever know it" (Holquist, 1990, p. 41). This idea is amazingly similar to the phenomenology of Merleau-Ponty. In this sense, the "I" is not the biological/physical object that occupies a given space, nor the mental self which is the ultimate source of existence (evident through the *cogito ergo sum*). It is a *sign* that points to a position (singularity) within the flux of being. Holquist explains this complex idea beautifully by saying that: "Much as Peter Pan's shadow is sewn to his body, the 'I' is the needle that stitches the abstraction of language [the sign 'I'] to the particularity of the lived experience" (1990, p. 28). Only through the sign "I" are we able to fix the singularity of being, made tangible by our flesh and its encounter with the world, into a definite object. Along the same line, Peirce presented the idea that man is a sign. According to Peirce: "When we think, then, we ourselves, as we are at the moment, appear as a sign" (1978, p. 169). Since our thinking is mediated by signs, even our conception of ourselves cannot escape a semiotic perspective. Indeed, it has been recently argued by Neuman (1999) that the incomprehensible Bakhtinian "I" is realized not only through language as a sign system, but also through certain artifacts/signs, such as the self-portrait, that stitch the incomprehensible "I" into concrete life.

Bakhtin's semiotic conception of the "I," which is process-oriented in nature, led him to the radical step of rejecting Aristotle's law of identity as concerns man:

> Man is never coincident with himself. The equation of identity A=A is inapplicable to him. In Dostoevsky's artistic thought, the genuine life of the personality is played out in the point of *non-intersection of man with himself*, at the point of his departure beyond the limits of all that he is in terms of the material being which can be spied out, defined and predetermined without his will, "at second hand." (1973, p. 48, my emphasis)

This statement suggests that not only "objects" do not really exist, and that signs do not obey the law of identity, even the "I," our secure anchor, is not an object, but a dynamic being sometimes fixed by the semiotic "I." Previously, we argued that the law of identity does not describe our primordial experience with singularities. Bakhtin makes the same point concerning man, and reminds us again that the law of identity is one of those value judgements we impose on a primordial experience in order to let the "ship" continue on its way (as suggested by Spencer-Brown). So how can we approach ourselves or others when all we have is an undefined flux of being demarcated and fixated by a social-semiotic system? Bakhtin's answer is that the genuine life of the personality can be approached only *dialogically*, and then only when it "mutually and voluntarily opens *itself*" (1973, p. 48). This unique conception of identity throws some new light on the meaning of being a self from a process-oriented perspective. However, it also calls for a further analysis of the relationship between the body and the mind. After all, our existence is grounded in flesh and in the uniqueness of the human body as our boundary with the environment. How can we speak of the mind and the self as abstract semiotic phenomena and, at the same time, locate them in the material existence of the body? The next chapter aims into inquire this issue.

Chapter 15
On Turing's Carnal Error

Summary: The analysis presented so far points at the body as an ensemble of boundaries and as the ultimate source of existence. The current chapter aims to elaborate upon this idea and to remind us again and again that: "The body is the vehicle of being-in-the-world, and having a body is, for a living creature, to be intervolved in a definite environment, to identify oneself with certain projects and to be continually committed to them" (Merleau-Ponty, 1962, p. 82).

A common conception, both among scholars and lay people of the Western world, is that the human mind is "body-free" and indifferent to the physical realm in which it is instantiated. As argued by Lakoff and Johnson (1999): "Much of the Western philosophical tradition assumes a form of faculty psychology, according to which we have a faculty of reason separate from our faculties of perception and bodily movement" (p. 37). Orthodox cognitive science has stood firmly behind "faculty psychology" and the idea that the mind may be envisioned as an abstract computational machine. This idea is epitomized in Turing's (1950) seminal paper, which describes the mind as a computational machine that involves the representation and manipulation of symbols. Special attention should be given to the fact that Turing's machine is a theoretical one, a machine which is defined by its computational activity rather by any specific physical form – the "hardware." According to this conception: "Intelligence...is only incidentally embodied in the neuropsychology of the human brain, and what is essential about intelligence can be abstracted from that particular, albeit highly successful, substrate and embodied in an unknown range of alternative forms" (Suchman, 1987, p. 8).

The classical computational conception of the mind, which powerfully dominates orthodox cognitive science (Anderson, 1993; Newell, 1990; Simon, 1969; Thagard, 1996), has been an important theoretical premise in the study of the mind. However, this conception may be criticized on the grounds that the mind is not a Platonic entity indifferent to the carnal body in which it is embedded. Rather, it is a system which has been profoundly shaped by the totality of human experience to include the body and its practice.

The notion that the mind is embodied has been discussed from various perspectives (Maturana & Varela, 1972; McCulloch, 1970; Varela et al., 1993). One of the most elaborated of these conceptions has been presented by the cognitive linguist, George Lakoff, and the philosopher, Mark Johnson (Johnson, 1987; Lakoff & Johnson, 1980; 1999). Lakoff and Johnson start with the well-established presumption that the process of conceptualization is fundamental to our activity in the world. The cognitive mechanism underlying this conceptualization is comprised of "conceptual metaphors." As they argue: "It is hard to think of common subjective experience that is not conventionally conceptualized in terms of metaphor" (Lakoff & Johnson, 1999, p. 45). These conceptual metaphors are grounded in the subject's basic sensorimotor experience with the world, and mapped "across conceptual domains that structure our reasoning, our experience, and our everyday language" (Lakoff & Johnson, 1999, p. 47). Hence, our thinking is metaphorical in nature, and our metaphors are patterned on the way we sense the world and operate physically in it. A small thought experiment could illustrate this point. Imagine yourself trying to describe your relationship with your spouse. You may describe your spouse as "sweet," "understanding" and the relationship as "warm" and "close." In this case, all the concepts you use, "warm, "close" and even "understanding" (under-standing!), are metaphors from the sensorimotor world. Christopher Johnson's theory of conflation (quoted in Lakoff & Johnson, 1999) explains this phenomenon by suggesting that in our early development, sensorimotor experience is routinely conflated with non-sensorimotor experience. This conflation produces undifferentiated sensual-conceptual experience in childhood, and conceptual metaphors embedded in the sensorimotor experience, in adulthood.

The notion of conceptualization as a metaphorical process grounded in bodily experience presents a serious challenge to "faculty psychology." In this context an interesting question arises: If embodied metaphors are the foundation of our cognitive processes, what kind of metaphorical-conceptual systems develop in people who experience the world through different sensorimotor channels? If people who experience the world through different sensorimotor channels develop different conceptual systems, this would lend support to Lakoff and Johnson's theory.

An indirect response to the above question can be found in Vygotsky's socio-cultural psychology. In his "Fundamental Problems of Defectology," Vygoysky (1993) suggested that a child whose development is impeded by a defect is not simply a child less developed than his peers but a child "who has developed differently" (p. 30). Vygotsky does not elaborate this difference. However, it must be remembered that he considered the mind to be the synergetic product of three different paths of development: the phylogenetic, the ontogenetic and the cultural. In this context, it may be inferred that the "difference" speculated by Vygotsky points to a different way of thinking as the result of a different course of bodily development. This speculation supports Lakoff and Johnson, since it supports the theory that different sensual experience would result in different metaphorical-conceptual systems. A short story by H. G. Wells (1947) and the well-known case of Helen Keller (1928) may give us some insights into this notion.

15.1. IN THE COUNTRY OF THE BLIND

In his short story, "The Country of the Blind," H. G. Wells describes a man called Nunez who found himself by accident in an isolated valley populated exclusively by blind men. "For fourteen generations these people had been blind and cut off from all the seeing world; the names for all the things of sight had faded and changed; the story of the outer world was faded and changed to a child's story. Much of their imagination had shriveled with their eyes, and *they had made for themselves new imagination with their ever more sensitive ears and fingertips*" (p. 22, my emphasis). In Wells's story the different conceptual systems that evolved from the differences in their sensorimotor experiences are evident when Nunez tries to explain to the blind men from where he came: "Over the mountain I come," said Nunez, "out of the country beyond...where the city passes out of sight." "Sight?" muttered Pedro [one of the blind men]. "Sight?" (p. 19). The different conceptual metaphors resulted in enormous communication difficulties and caused the blind to question the meaning of Nunez's words, and to suggest that "He stumbles and talks unmeaning words" (p. 20), that "mean nothing with his speech" (p. 22). The native medical doctors examine Nunez and reach the scientific conclusion that his maladaptive behavior results from two mysterious balls (the eyeballs) that appear to be malignant tumors that press the brain. Their diagnosis leads them to suggest the removal of those tumors in order to bring Nunez back to normality ...

The blind men who were not able to sense the world through vision constructed a different psychological realm that excluded vision and the conceptual metaphors derived from it. Buerklen, who similarly speculated about this process (1924 quoted in Vygotsky, 1993), suggested that: "They [the blind] develop special features which we cannot observe among the seeing." However, Buerklen points to the cultural aspect of the mind by saying: "We must suppose that if *the blind associated only with the blind and had no dealings with the seeing,* then a special kind of people would come into being" (p. 34, my emphasis). Buerklen's suggestion is important since it analyzes the development of the mind not only in the context of the carnal body and its development, but also in its proper cultural context. This point is missing from Lakoff and Johnson's theory, a shortcoming that will be discussed further below.

The case of Helen Keller (Keller, 1928) gives strong support to the insightful literary imagination of Wells. Helen Keller became blind and deaf in her early childhood. She communicated with the world through smell and contact and learned the use of language in an older age through the assistance of her legendary teacher, Miss Anne Sullivan. Although Keller's autobiography is not typical scientific evidence, much of her introspection supports Lakoff and Johnson, Vygotsky and the previous insights presented by Wells. In her book, Keller considers the relation between language and thought. She notes that prior to learning language she was not able to think in concepts. In her introspection, Keller questions the meaning of thought without language by saying: "This thought, if a wordless sensation may be called a thought..." (p. 23). In this sense, the concepts we use to think with are the essence of our mind and give rise to its activity. After learning a language, "everything had a name, and each name gave birth to a new thought" (p. 24). This evidence supports the Vygotskian notion that thought is realized through language, and Lakoff and Johnson's notion of the embodied mind: Since Keller was not able to experience the world through the most important senses of sight and sound, her ability to develop a metaphorical-conceptual system with which to think was severely impeded.

Wells's story and Keller's case support the idea that the mind is part of the body and that different bodily experience may result in different minds. In this sense, these cases clearly support Lakoff and Johnson's theory of embodiment. However, the cases I described have been *deliberately misleading*. Because they are not common cases, but unique cases representing closed systems (Keller's mind and the isolated valley populated exclusively by blind men), they isolate the effect of the body on the development of the mind from a wider context. In contrast to Keller's and the blind population's isolated worlds, most of us live in an open dynamic system, a world inhabited by different others who interact, evolve and change in an enormous diversity across history, geography and society. It is in this context of flowing and divergent existence that the notion of embodiment should be discussed. This is also the background for my critique of Lakoff and Johnson's sense of embodiment.

15.2. MIND WITHOUT CULTURE

Lakoff and Johnson present a powerful conception of embodiment, presented as an antithesis to faculty psychology and the Cartesian tradition. However, as proponents of cognitive-science, it seems that they cannot avoid Cartesian elements that underlie their cognitive discourse. In this section, I present three major difficulties associated with Lakoff and Johnson's embodied mind. The first difficulty concerns the indifference of the embodied mind to the cultural system in which it is located. The second concerns the conception of the body as a pre-existing physical-material object, which determines the mind in a linear and mechanistic sense. The third difficulty entails the second difficulty: If the body determines the mind in a causal sense of the term, then we should ontologically assume the existence of two different entities, the mind and the body, a Cartesian dualism Lakoff and Johnson seek to avoid.

The idea that the human mind is not separable from bodily experience can be traced to Merleau-Ponty (1962). In general, the phenomenologists ignored the cultural aspects of the human mind (Turner, 1996), although Merleau-Ponty was trying to embed his phenomenology in cultural situations through a reliance on Claude Levi-Strauss (Priest,

1998). Lakoff and Johnson praise the intellectual heritage of Merleau-Ponty but ignore his holistic notion of embodiment and present a form of embodiment which is indifferent to the cultural system. This critique has been already raised by Sampson (1996), arguing that *"history, culture and community are excluded* [from Lakoff and Johnson's embodied mind] *even as the body has been added"* (p. 619). In this sense, our bodily experience seems to be the determinant, in the most mechanistic and individualistic sense of the term, of our conceptual system and the role of culture seems to be minor: Whether you are a Chinese Buddhist in the tenth century, a Catholic Italian in the seventeenth century or an ultra-orthodox East European Jew in the fourteenth century means nothing to your embodied mind. In all of these cases one's basic and universal human physical structure will be the most [or the only] important determinant of one's mind. This notion ignores Buerklen's socio-cultural comment previously mentioned and Merleau-Ponty's holistic view that each aspect of being-in-the-world, including being-in-the-cultural-matrix-of-the-world, is an "indispensable moment of the lived dialectic" (Merleau-Ponty, 1965, p. 189) and cannot be reduced or discussed in isolation.

Lakoff and Johnson support their thesis of the primacy of the body over other sources of metaphors by presenting a case for the universality of these embodied metaphors. For example, one of the central schematic structures Lakoff and Johnson present is the *containment* schema of within vs. without. This "in-out" metaphor seems to be a powerful explanatory mechanism since it uncovers the embodied nature of major structures that underlie Western thought, for example, the binary structure that underlies Aristotelian logic and other aspects of Western thought (Kilgour, 1990). Surprisingly or not, the binary matrix that characterizes the Western world is not a universal psychological structure of the human mind, and different forms of logic have evolved in different socio-cultural contexts. Aristotelian binary logic, which has been considered for centuries as manifesting a divine truth, has been shown to be a specific cultural construct, its universality being questioned by recent developments in fuzzy and multi-valued logic (Kosko, 1993).

If different socio-cultural contexts can result is different forms of logic, and if different forms of logic can evolve without a radical change in bodily experience, the simple structuralist agenda of Lakoff and Johnson, with its anchor in a universal physical body, needs to be qualified. In other words, the development of different conceptual metaphorical systems clearly calls into question the claim that embodied metaphors are primary in all human consciousness.

The second worrisome issue in Lakoff and Johnson's theory is that there seems to be only a one way flow of influence from our bodily experience to the metaphorical-conceptual system of the mind: The body is an active force that determines the mind in a causal way, and the mind becomes a passive "dependent variable" without any influence on the body that determines it. As Lakoff and Johnson (1999) suggest: "What is important is that the peculiar nature of our bodies *shapes* our every possibility for conceptualization and categorization" (p. 19, my emphasis). This notion is clearly located in a positivist-Newtonian ontology (Harre & Gillett, 1995). It is faulty because it ignores the cybernetic nature of the mind (Bateson, 1973; von Foerster, 1974), and therefore the possibility that the mind may recursively transform itself and transcend its concrete embodiment. Lakoff and Johnson's causal conception of embodiment fails to explain human creativity and the changeable nature of the human mind in different socio-cultural contexts. In fact, by merging socio-cultural with cybernetic notions of the mind, it may be argued that the embodied mind can recursively change its "body" and therefore itself. For example, by creating new forms of sensorimotor experience, through technological artifacts such as bionics or advanced surgery, the human mind can change its bodily experiences, create a whole new spectrum of metaphors and, indeed, transcend itself.

The third problem is a derivative of the second. Lakoff and Johnson rebel against Cartesian dualism by placing the mind in the "great chain of being." However, by pointing to a one-way causal relationship from the body to the mind, they have fallen precisely into the Cartesian dualism they were trying to avoid: They have to assume that the mind and the body are two different and distinct *entities,* and that the *res extensa* (the body), which is a universal structure, determines the structure of the *res cogitans* (the mind). This problem results from Lakoff and Johnson's realistic conception of the body as a self-evident physical entity. After all, who can deny the fact that human beings are featherless creatures with two legs, two hands, and fingers and thumbs? My intention is not to deny the reality of certain singularities we call "bodies," but to adopt a contextual-cybernetic conception of the mind which suggests that these "structures" do not exist pre-reflectively, nor do they have any meaning separate from the observing minds that conceived and constructed them in different cultural contexts. In this sense, and as has been suggested by Merleau-Ponty, the human being is a complex unity that cannot be reduced to its components. The idea that mind and body are "organizational characteristics" (Bateson & Bateson, 1988) rather than two different entities will be elaborated further below.

The most basic distinction-boundary exists between the organism and everything else. This distinction is what constitutes the identity of the organism, whether human or non-human. In this sense, a human being exists as a unity which is constituted by the relation of conglomerated components: the observer, the environment, the difference between the two and the emerging whole. Such a contextual approach may suggest *that our mind is the totality of boundary relations rather than a given substance within our skull.*

Adopting this approach we may consider the body not as a simple object preceding our mind, but as a major source of boundary relations that constitutes mind as a boundary phenomenon, as mind itself. Therefore, the basic boundary between the subject and everything else may be based on bodily experience, as suggested by Lakoff and Johnson. However, it must be noted that the body is only one source of boundary and that the differences that constitute our mind may be triggered by other boundary relations, such as cultural boundaries as well.

In order to explain this argument it is useful to think of the boundary as a context: "a set of observed relation between an environment and that which appears to be contained within a distinction seen to be embedded in that environment" (Rebitzer & Rebitzer quoted in Von Foerster, 1974, p. 454). A context, as a set of relations, may be defined not only according to body-environment boundaries, but also through other boundary relations set for the individual. In this sense, our bodily experience is only one of several contexts that determines the forms of mind; not as the mechanistic *cause* of the mind but as one of the major sources of meaning. Since the mind is a multi-layered phenomenon and exists at different levels of logical analysis, to include the cultural level of analysis, it is certainly legitimate to explore the role of culture in constituting the boundaries of the mind.

Cole defines culture as "the entire pool of artifacts accumulated by the social group in the course of its historical experience" (Cole, 1996, p. 110). In the interactionist context of the Batesonian mind, it is useful to think of artifacts not simply as mere material entities, but, rather, as parts of the interacting components (boundary relations) that constitute the mind (Neuman & Bekerman, 2000). In this sense, it is better to use the term "artifactual,"[40] a term that emphasizes the dynamic nature of human products and the emergence of the mental processes through the interaction with artifacts to include signs as artifacts. Following this line of reasoning we cannot consider artifactuals, such as information technology, as simple projection of bodily experience, since the artifactuals exist on a different level of logical analysis. However, it can be argued that if the mind is a multi-layered system constituted by boundaries, there is room to inquire into the process by which artifactuals define the activity of the mind.

Considering the mind as a structure of boundaries that exists at different levels of logical analysis not only suggests a wider perspective on embodiment, but also addresses some of the difficulties associated with traditional theories of the mind. For example, Lakoff and Johnson totally ignore the recursive nature of the mind and its ability to redefine its boundaries. This avoidance can be explained by the enormous difficulties the notion of self-reference has raised in the grand Western traditions: the rationalist and the empiricist. The idea that the mind is a boundary phenomenon solves the difficulties associated with self-reference by locating the Archimedes point of self-reference in the artifactuals (e.g., signs) that re-constitute the boundaries between the subject and his environment rather than inside the mind as a closed system (as proponents of the rationalist tradition suggest) or in the external world per se (as proponents of the empiricist tradition suggest). This suggestion has some radical implications, since it suggests that the mind may transform itself through artifactuals (Luria & Vygotsky, 1992; Neuman & Bekerman, 2000) and that the emergence of artifactuals, at the cultural level of analysis, may result in new psychic phenomena, such as consciousness we usually associate with the individual level of analysis (Neuman, 1999).

15.3. EMBODIMENT AND MIND

The relation between bodily experience and the mind has been elaborated in several theories of embodiment, such those of Piaget (1971), Merleau-Ponty (1962), and Lakoff and Johnson (1999). Previously I rejected the idea that the mind may simply be reduced to bodily experience. I also argued that we might consider the body as a distinction or difference, which is determined as the interface between the organism and its environment. In other words, and from a psychological-developmental point of view, the basic boundary between the organism and everything else is based on bodily experience. However, it must be noted that the body is *only one source* of boundary definition and that the differences that constitute our mind are constituted by other boundary relations as well.

For example, an interesting discussion concerning the way bodily experience shapes our mind appears in the work of the great French mathematician, Henri Poincare (1952). Poincare was bothered by the question: What is the source of our spatial experience? He argues that "we could not have constructed space if we had not had an instrument for measuring it" and that the "instrument to which we refer everything, which we use instinctively, is our own body. It is in reference to our own body that we locate exterior objects, and the only special relations of these objects that we can picture to ourselves are their relations with our body. It is our body that serves us, so to speak, as a system of axes of co-ordinates" (p. 100). This idea was adopted and developed by Piaget in his theory of genetic epistemology. For example, in his seminal work, "Psychology and Epistemology," Piaget investigates the phenomenon known as "object-permanence." He suggests that "permanence of the object is thus closely linked with its localization in space and, as we see, localization itself depends on the construction of the 'shifting group,' which Poincare rightly established as the development source of sensorimotor space" (Piaget, 1971, p. 16). From a more general perspective, Piaget suggests that: "We only know an object by acting on it and transforming it...And thus there are two ways of transforming the object we wish to know. One consists of modifying its position, its movements, or its characteristics in order to explore its nature: this action is known as "physical." The other consists of enriching the object with characteristics or new relationships while retaining its previous characteristics or relationships, yet completes them by systems of classification, numerical order, measure and so forth: these actions are known as "logico-mathematical" (p. 67).

Piaget failed to explain the way "physical" actions we perform on objects (or more generally our "sensorimotor experience") are related to the "logico-mathematical" actions and, therefore, was criticized for being a secret Platonist who believes in innate structures that govern our mind (Piatelli-Palmarini, 1980).

The mysterious relation between "physical" and "logical" operations – the embodiment of the mind – may be clarified if we accept the idea that we cannot consider our mind as a simple projection, transformation or representation of bodily experience with concrete and predetermined material objects, but rather as the boundary relations that exist as *correspondences* between the observing and an observed system. Although our body and its operations on "physical" objects are the basic sources of our mind, our mind consists only of abstract patterns that correspond to the correspondences between the observing and the observed system. The ability to compute these correspondences (to behave according to a difference that makes a difference) is what distinguishes between living organisms and non-living systems, and the ability to compute these correspondences at the meta- (semiotic) level is what distinguishes between human beings and organisms that are captivated by their concrete operational interaction with the observed system. This position clearly transcends the classic rationalist and empiricist paradigms. However, it is also a radical shift from classic structuralism and constructivism as represented by Piaget.

What we may learn so far is that our mind is the synergetic appearance of socio-somatic-semiotic threads. The idea of a multi-level, multi-distinction mind which operates according to a self-referential and, therefore, non-linear dynamic in a multi-agent (social) environment, calls for a unique perspective for inquiring into its nature. Complexity sciences may be a good candidate for this agenda. What is complexity and how we may use its radical theoretical arsenal, in the next chapter.

Chapter 16
What is so Complex about Complexity?

I have yet to see any problem, however complicated, which, when you look at it the right way did not become more complicated.
Poul Anderson

> Summary: Complex systems are systems characterized by a large number of interconnected components. Those systems resist simple reductionism and, therefore, the researcher should study the overall dynamics of a complex system as evident from the emergence of gestalt forms out of micro-level interactions. Should we examine the mind as a complex socio-somatic-semiotic system?

So far, we have developed the thesis that our phenomenal world emerges as an expansion of the primary distinction. Each one of us is a distinction who constitutes his/her/its existence through the recursive and hierarchical dynamic. We also learned that distinctions are evident through indications, signs, the names we give to the distinctions, and that those signs are indeterminate creatures that have meaning only in context, in a spatial-temporal event of interaction between agents. The context that gives meaning to our signs is social. The problem is that, on the one hand, our existence as minds is dependent on our boundary structure that constitutes the systemic closure of each of us. On the other hand, the mind is a social-semiotic system. So how is it possible to bridge the gap between the systemic closure of the primary distinction with its bodily basis and the social realm? The question may be rephrased as how do we deal with a context in which numerous agents operate in a way that generates metastructures (the social semiotic system) that in return gives meaning to the individual agents who produce it and constitute their systemic closure. A possible direction is to inquire into this system by adopting a complex system's perspective. This chapter introduces the idea of complex systems.

Classic science as it appears to us has celebrated a reductionist approach to the study of natural phenomena to include celestial and carnal bodies alike. The idea behind reductionism is apparently very simple. The researcher encounters a phenomenon or phenomena (s)he would like to explain. Explanation here means the ability to represent a perplexity of appearance by formulating it in a clear and simple (mathematical?) language that can be used beyond the specific phenomenon. In order to accomplish her task she breaks the phenomena under inquiry into well-defined (and usually, but not necessarily, homogenous), small numbers of components. Those components are the building blocks of her theory. After breaking the phenomenon into small pieces, which are considered to be crucial to the model building, the researcher tries to formulate the rule/s that connect those components, and through those rules to predict the behavior of the system under inquiry. This venture emphasizes the isolation of components (variables) and the study of the system as a whole through the assembly of a *few isolated components*. The excitement surrounding the Genome Project epitomizes the classic reductionist venture. We were promised that within the near future, and through the computation power of both computers and researchers in bio-informatics, the "book of nature" would be disclosed, complex behavior would be reduced to a certain gene/s and the mystery surrounding life would be expelled. Thrill seeker? Yes, of course, we will identify the specific gene. Introvert or extrovert? Very simple. There is a specific gene that can easily explain your character. Aha! you prefer young, blonde models (Pamela Anderson style) to dark haired models (Naomi Campbell style)? We, the scientists, will find the naughty gene responsible for your sexual preference. Previously, I presented an excerpt from Barth's book that mocks the imperialist pretensions of the Freudian reductionist venture. This excerpt clearly applies to the Genome Project. Whether you try to reduce complex phenomena to the Phallus or to the Genome, some systems clearly resist this form of analysis.

The above argument should be qualified. Reductionism has been shown to be productive in many ways. However, the Genome Project celebrates not only the high point of reductionist venture, but also its ultimate point of decline because some members of the scientific community start to realize that many mysteries surrounding life cannot be solved by simply breaking them down into a few small pieces (e.g., Cohen, 2000), and extensive work should be done to understand how to put the pieces together.

The general problem with reductionism is that it tries to explain one level of a multilevel system by a lower order level without acknowledging the dynamic that constitutes those levels as *separate* levels of analysis. Consider for example the economy of a city. Can it be explained by reducing it to the behavior of the individual consumer or the simple aggregate of the consumer behavior? The answer is definitely not, as evident from the failure of this "Robinson Crusoe" theory in economics. What about the behavior of a group? Can we explain the behavior of a group by studying the aggregate behavior of the individuals that compose the group? Anyone who knows something about group dynamics will dismiss this reductionist attempt. Don't worry, there is an alternative to hard-headed reductionism.

Indeed, some researchers who realize the limitations of reductionism do not blame the environment for playing a hard game, but look for different ways of conceptualizing these systems. These systems are described as "complex" systems. We should recall that the term "complexity" comes from the Latin "complexus," which means "containing." Something is complex if it contains something(s). Like the

Russian dolls, complex phenomena contain something within. Metaphorically speaking, we should insert our head into the "whole" in order to see what it contains, rather than breaking this whole into smaller and smaller pieces. A good example of a complex system is the game of chess. Chess is a game with a small number of well-defined components. Those components operate according to given rules that constrain their movement. In contrast to real kings, the king in a game of chess is forbidden to move wherever he would like to. Surprisingly, a limited number of components and a few rules of interaction generate an enormous number of states. This is what makes this game so attractive, in contrast to simple games such as tic-tac-toe. The same is true for natural language. Natural language, with a finite set of components (signs) and syntax that constrain the arrangements of those components, is able to generate enormous quantities of sentences and other macro-structures, i.e., natural language is a complex phenomenon.

One of the essential terms associated with the study of complex systems is *emergence*. Emergence can be defined as, "the emergence of higher levels of integration through the functional association of heterogeneous elements of lower complexity" (Francois, 1997). This term calls our attention to the fact that there are systems in which the phenomenon under inquiry "emerges" or "pops" out of *interactions* between the components of the system. That is, the system is reduced to basic components, but the focus of our inquiry is on the overall dynamic between the components that give rise to a unique form of appearance that cannot be derived from a few basic components or their simple aggregate. The whole is different from its parts, more accurately, from the interaction of its parts, and complexity is the general term we use for a variety of ideas that try to deal with the dynamic governing the emergence and the behavior of such wholes.

It is important to note that this overall dynamic unfolds in time. In contrast with static models of representation, such as classical logic, the modeling of a complex system should involve a dynamic representation that shows how the system moves between different states, and how this movement constitutes the appearance of the whole we are familiar with. There is no embryology without studying morphogenesis (the creation of morphemes) and the states the embryo goes through, fertilized egg to mature adult. There is no developmental psychology without identifying the dynamics and the transformations of mental states from infancy to maturity and, of course, there is no semiotics that can avoid the emergence of the symbol out of the primary distinction.

The term emergence is associated with a non-reductionist movement, but it should not be confused with another non-reductionist movement – Vitalism. Vitalism was a number of anti-reductionist approaches that shared the idea that life may be traced to some kind of mysterious and non-material spirit. Henry Bergson, we previously mentioned, was one of the main spokespersons of this movement. Today, biologists are rarely occupied with the most important property of their main research subjects – life. Life is a concept that cannot be explained in material terms or through other terms that biologist, physicists or chemists are familiar with. This mystery does not turn most biologists into vitalists. On the contrary. Most biologists consider Vitalism to be an anti-scientific obscurity and not as a concept of any worth. Therefore, they prefer to deal with biological systems through the "materialistic" tools and to leave the question of life and its origins to philosophers and other harmless and obscure creatures. There is a famous anecdote about Descartes: Descartes once came to his mother and told her that he would like to be a mathematician because a mathematician needs only a pencil, a table and a garbage can in which to throw his unsuccessful meditations. Several years later, Descartes turned to his mother and told her that he had decided to be a philosopher because a philosopher does not need the garbage can! This anecdote was told to me by a well-known computer scientist and the message is clear. In philosophy anything goes. Therefore, philosophers, semioticans, poets, historians and other scholars who do not need the garbage can are left to deal with the concept of life, and the serious people (those with the white coats and the garbage cans) do science. Real science. Well, the failure of reductionism in science to confront the complexity of many real world phenomena is the same as the failure of Vitalism in science. While reductionism tries (and fails) to explain many phenomena by reducing them to basic physical components, Vitalism tries (and fails) to explain almost everything by reducing it to something that cannot be explained in material (physical) terms.

The anti-reductionist venture epitomized in the writings of Bergson and Whitehead should be located in a general intellectual atmosphere that existed in Europe at the beginning of the twentieth century (Harrington, 1996). In contrast with what they portrayed as the fragmented, mechanistic and dead universe of the Newtonian physics, some of the leading holistic figures in Germany sought holistic and spiritual science. For example, Goethe imagined a "rich and colorful world shaped by aesthetic principles of order and patterning" (p. 5). This vision can be easily portrayed as anti-scientific or as a return to "black occultism" [a term used by Freud in order to describe Jung's psychology] of the Dark Ages. However, Goethe sought to explain the world in terms of "a small number of fundamental forms or Gestalten" by "observing and comparing the various metamorphoses of one or another form, he felt that the original or primal form of the type in question could be deduced" (p. 5). This vision is strikingly similar to the vision portrayed by the proponents of complexity science with its emphasis on emerging wholes, morphogenesis and various kinds of attractors! Considering the holistic venture in terms of complexity science may provide us with a non-linear and liberating perspective on the history of science. History, rather than crystallizing into a superb mind, like Hegel thought, is non-linear, and up-to-date advances in science can be easily traced to ancient ventures that

What is so Complex about Complexity? 129

have been easily dismissed. By that I do not mean that Goethe's vision of science should be fully or easily embraced. Goethe's conception of science is imbued with a theological and teleological spirit that does not find its proper place even in the modern science of complexity. However, there is an important point in his suggestion to seek wholes rather than components.

The roots of complexity science are also evident in gestalt psychology, a movement which sought the general forms we impose on our environment. The link from the gestalt psychologist, Max Wertheimer, to Benedict (in Hebrew Baruch) Spinoza is particularly of interest since both Goethe and Wertheimer acknowledged their intellectual debt to Spinoza, who may be considered the "Godfather" of complexity science. Indeed, Spinoza may be considered the father of self-organizing systems[41] and his influence on some of the leading figures of holism, such as Goethe and Wertheimer, is well documented. The next dialogue with the cat aims to explore further the relations between Spinoza's philosophy and the venture of complexity sciences.

Cat-logue 8
Speaking with the Cat about Spinoza

Bamba: You are surprising me again. Beyond racial issues, I never thought that Spinoza had anything to do with complexity sciences.
Dr. N: What do you mean by "racial issues"?
Bamba: Well ... I do not want to be portrayed as a racist ... but in a recent book about complexity by Levine, I read that there is a prominent English biologist who argues that, while the leaders of evolutionary theory are English country boys, urban Jews are among the leaders of complexity sciences...
Dr. N: Sounds racial and I'm not interested in commenting on it. Anyway, I didn't know that you are reading books on complexity.
Bamba: I am a hunter and, as a hunter, I have plenty of time to read. I don't have to eat all the time like a cow, or worse, like a human being who eats and watches television at the same time.
Dr. N: Quite impressive and to think that Rene Descartes considered animals the same as machines – without spirit.
Bamba: Since you mention this horrible philosopher, it is worth noting that Descartes's system is clearly a dualistic system that differentiates between body and mind, God and the world. Indeed, only by assuming such a sharp differentiation between body and mind was he able to preserve his cherished self as contrasted to simplistic materialistic conceptions that reduce the human spirit to simple matter.
Dr. N: Well, today we know that there is no such thing as simple matter. However, you are right and you can see that Descartes reacted to the same materialistic and reductionist tendencies the German holism reacted at the beginning of the twentieth century. Unfortunately, Descartes's anti-reductionist venture ended in another form of reductionism, an upward kind of reductionism to the immaterial self, which is not less problematic than its materialistic counterpart. Spinoza's philosophical system is a reaction against this dualism.
Bamba: Indeed, Spinoza does not consider the self as its starting point, but rather starts with God, with the ultimate and eternal substance, with what you have described before as nothingness.
Dr. N: Right. Although in everyday talk nothingness is generally associated with something negative, Spinoza adheres to the ancient mystical conception of nothingness as equal to the ultimate whole. In Spinoza's philosophy, God is the ultimate and complete being and, therefore, it is described as *causa sui* – the cause of itself.
Bamba: That's quite similar to Aristotle's definition of God and metaphysics.
Dr. N: Right. The radical implication of this conception is that the material world is part of God and, therefore, the reasons for its behavior should be sought inside the system. What you see is what you get. So the

interesting thing that is derived from this conception is that explanatory mechanisms outside the system are rejected. No more Platonic ideas from the outside. No more transcendence.

Bamba: Amazing! According to this logic, I am a part of God himself!

Dr. N: Your conclusion was an enormous problem for philosophers from the Jewish and Christian tradition who consider God as transcending the world rather than as immanent in the world and its beings. And you are right, Spinoza is saying that every being is a *modus* of God. Every living creature is a part of the general logos. However, please remember that God according to Spinoza is not the old man with a white beard portrayed by Michelangelo. God is the logic of being. God is the ultimate and necessary cause of being, not in his existence outside being, but by the fact that beings are a logical part of this logic, *the same as the whole is a necessary cause of its parts*.

Bamba: My goodness, this is exactly the idea underlying gestalt psychology, holistic thinking and complexity science! Are we getting closer and closer to the logic of distinction and the dynamic that generates the primary distinction and the world of phenomena?

Dr. N: Correct. Just look at two important aspects of Spinoza's philosophy: (1) the idea that the reason of the phenomena should be sought *inside* the beings and not outside them and (2) that individual beings are always a part of the ultimate *whole*. These aspects clearly draw the connection between Spinoza and the logic of distinction.

Bamba: Would you like to close this discussion by saying something about Spinoza's epistemology?

Dr. N: I just want to mention that Spinoza differentiated among several levels in which we know the world, the highest of them all being intuition.

Bamba: Do you mean feminine intuition?

Dr. N: Not exactly, although my wife strongly insists that intuition is a female property. By intuition, Spinoza means an unmediated knowledge of reality.

Bamba: Hey, we spoke about intuition earlier in this book. I can recall that Peirce objected to this kind of knowledge.

Dr. N: Indeed, but other people, such as Henry Bergson and Wilhelm Dilthey, developed intuition as their primary methodology of scientific inquiry. Just remember that Spinoza conceives intuition as *an unmediated knowledge of the parts within the ultimate whole of being*. Spinoza is even speaking about intuition in emotional terms of the individual soul that recognizes its proper place in the totality of being. In this context, you should recall the Jewish sages who speak about creation ex nihilo, Spencer-Brown and his laws of form, and Bakhtin's holistic and dynamic conception of the "I." In addition, you should think about the fact that although our thinking is mediated by signs, only our ability to grasp the totality of those signs-in-use is what constitutes our understanding. That is, intuition is what exists in-between. It is the logic that constitutes the phenomenology of boundaries we previously presented.

Bamba: A final question.

Dr. N: Sure.

Bamba: I have a question that concerns Spinoza. According to Spinoza all forms of being take part in the ultimate whole. According to this conception, and in contrast with Descartes, I am not inferior to you. In this context, my question is as follows: "If my ethical status is equal to yours, which implies we have the same rights in this apartment, can I sit in your favorite armchair?"

Dr. N: In my favorite armchair?! Definitely not!

Bamba: But how can you refuse my humble request after reading Spinoza?

Dr. N: That is a good question, but remember that I treated you nicely after reading Descartes and that next week I am going to read one of the scholarly philosophers who wrote some nasty things about animals. Are you sure that you would like me to act according to the philosophers I read?

Bamba: Definitely not. But can I sit on your lap while you sit in the armchair?

Dr. N: Sure, come on and let's read together an interesting book written by Deleuze. Do you know that he was a great admirer of Spinoza ...

16.1. EMERGENCE AND SELF-ORGANIZATION

After introducing the general notion of complexity and some of its historical origins, we may turn to a deeper understanding of the terminology underlying complexity. Complex systems are dynamic systems. They are systems that change in time. There are two major components for such a system. First, the space in which this dynamic takes place, what is usually called the *manifold*, and second, the mathematical rule that describes the dynamics of the system, a *vector field*. What we usually see when we represent a complex system is a path the system follows from its initial state. This path is called the *trajectory* and the end of the trajectory is what we call *attractor*. An attractor is "a limited region of the phase space (i.e., a point or a set of points) toward which the trajectory of the system converges, tending to a steady state or periodic motion" (Francois, 1997, p. 31). In other words, an attractor is some kind of a stable state into which the system settles after wandering for awhile. Although people's movement in a shopping center may look arbitrary, the careful observer may notice that some of the smaller people (i.e., children) who visit the center are attracted by the candy stores. The random movement of the children settles into a relatively steady state of wandering around the candy stores and asking their mothers (fathers are less responsive) to buy the attractive sweets. Systems do not necessarily settle in a stable state. Some systems wander around forever and we, as observers, cannot recognize the attractors or the order in their long-term behavior.

There are different kinds of attractors. Unfortunately, death is a *fixed-point attractor*, although some religions suggest it is not. A sign may be considered a fixed-point attractor ("a pipe is a pipe") if we adopt a naive referential conception of meaning. An attractor can also be *oscillatory*. Think about paradoxes. Paradoxes express an oscillation between values of true and false, but you will never catch them setting peacefully on one position. They never settle on true or false. For example, is the statement, "This statement is false" false? If it is, then it is not. If it is not, then it is. A paradox is a disturbing phenomenon for those who are looking only for fixed-point attractors in living systems and forget that the source of being and meaning is a paradoxical movement evident in the behavior of the boundary. In life only death depicts the ultimate fixed-point attractor, while life is a paradoxical and oscillatory process. Another type of attractor is a *limit cycle attractor* that expresses a periodic orbit or a cycle of states. Although a weather system is considered to be in general a chaotic system, in some countries summer always follows spring that follows winter that follows fall. The strangest attractor of all is called a *strange attractor*. Here we get into the area of Chaos researchers. This kind of attractor characterizes large interconnected, deterministic and non-linear systems that are extremely sensitive to initial conditions of the system. Even a slight difference in the system's basic state would lead to enormous differences in the long-run. If one examines two points of an attractor and then follows their trajectories, one may find that the trajectories obtained are very different. Since the term "chaos" has become fashionable, it is worth dwelling a little bit more on its exact meaning. The term chaos is used in a mathematical sense to describe a family of mathematical mappings that appear random, but which have an underlying deterministic dynamic (order). Let us illustrate this general phenomenon by using one specific form of mapping known as *Arnold's Cat Map*. Before presenting this specific map, we should introduce the term *mod*. If we have a real number x, then the expression *x mod 1* denotes a number in the interval (0, 1) that differs from x by an integer. For example, 2.3 mod 1 = 0.3 because the number 0.3 differs from 2.3 by the integer 2. Arnold's Cat Map is a transformation that involves the mod operation:

$$(x, y) \rightarrow (x + y, x + 2y) \bmod 1$$

This means that given two real numbers x and y, our mapping transforms them into (x + y, x + 2y) mod 1. Now let us take a square in which the location of each point may be indicated by the pair of coordinates: x and y. We randomly chose a point in this square and operate on the coordinates of this point through Arnold's mapping. After applying this operation repeatedly, that is, after using a large number of iterations, we may see that our square is homogeneously covered by points. The points do not cluster in any particular region of the square, but rather spread in what seems to be governed by a random process. In addition, if we will follow the trajectory of two points that were located at the same coordinate, we see them diverging as time unfolds. Here we get into one definition of chaos. We say that a set D of points in S is *dense in S* if every circle centered at any point of S encloses points of D no matter how small the radius of the circle is taken. By using this terminology we may define chaos as follows: "A continuous mapping T of S onto itself is said to be chaotic if: (1) S contains a dense set of periodic points of the mapping T, and (2) there is a point in S whose iterates under T are dense in S" (Anton & Rorres, 2000, p. 687). There is nothing like a good mathematical definition to take our breath away, but at the same time to point at the lack of its relevance for inquiring into the human mind. Are you familiar with a single human behavior that can be described through this mathematical mapping? So, from now on we should avoid using the mathematical definition and settle for the general idea of chaos as a deterministic and non-linear system which is extremely sensitive to initial conditions and appears to us to be random.

Let us get back to the general idea of complex systems and demonstrate the idea of complexity by using networks. Assume that you have a simple system that is composed out of three units/nodes (the holy triple again) and connections between the nodes. To simplify matters, let us assume that each unit can have only one of two values, either 1 or 0. This is not a necessary assumption since nodes can have values ranging from 0 to 1, or any other possible value, but then the situation becomes much more complicated. Our network is composed of the three binary nodes and each node signals to the other two nodes. It can easily be shown that each node can receive four possible inputs from the other nodes: 00, 01, 10 and 11. Now we should choose a rule in which the values of the inputs will be combined. Let us choose the logical operator "AND." If we apply this rule, there are several possible outcomes. If the two nodes equal 0, then the output is "0 AND 0 = 0." If the two nodes equal 1, then "1 AND 1 = 1." If one of the nodes is 0 and the other is 1, then "0 AND 1 = 0." Note that the rules of logic are different from the rules of arithmetic you were taught when you were a schoolchild. In arithmetic "1 + 0 = 1." However, in Boolean Logic, where 1 symbolizes "True" and 0 symbolizes "False," the outcome of A "AND" B is true only if the two components are true. In sum, we determine the components of the system, their number, the possible values of the components, the number of connections between the components and the rule(s) governing those connections. If we examine this network closely enough we can see that there are 8 possible states the system can reach: 000, 111, 001, 010, 100, 011, 101, 110. This is the space in which the dynamic takes place, the *manifold* of the system. The next step is to examine the dynamic of our system. We can do so by randomly assigning values to the units and by letting the system run. If we assign the following values to the system at T0: node 1 = 1, node 2 = 1 and node 3 = 0, then at T1 the outcome is:

node 1 = 0, node 2 = 0, and node 3 = 1

And at T2 the system will look like this:

node 1 = 0, node 2 = 0, and node 3 = 0

Following the trajectory of the system to T3 we find that:

node = 0, node 2 = 0 and node 3 = 0

After going through a few cycles for a bit, our system has peacefully settled or reached an attractor. This description looks simple because after only a few cycles the system settles. However, try to figure out what happens when the number of units/connections in the system is larger. If we have 30 units connected to each other, then there are 2^{30} possible states. If we have 300 connected units, then there are 2^{300} possible states and so on. In other words, the increasing number of units and connections eventually leads to a situation in which it is impossible to trace the whole state cycle of the system. However, several trajectories may lead to the same state cycle or attractor. These trajectories, paths that lead to the same state cycle, are called the *basin of attraction*. It is as if the state cycle absorbs the different trajectories into the same place.

The meaning of emergence is closely associated with the dynamic of attractors, since the state cycles we recognize are relatively stable patterns that emerge out of local interactions between the components of a system. No one forces them on the system. Those attractors are the result of the systems' internal dynamics. In this context, we should introduce the concept of *self-organization*. Due the enormous popularity of "complexity research," the term "self-organization" has become so widely used and in so many senses that it is impossible to present a single comprehensive definition of it. I believe that the main idea behind "self-organization" is that the system under inquiry behaves in a way we cannot fully explain from outside, but through the system unique dynamic as a systemic closure. For example, our ability to maintain our body temperature when facing perturbation from the environment should be examined through the biological feedback system of our body. Our unique bodily configuration should be studied through the unique dynamic that transforms the fertilized egg into the unique configuration of the adult. Our unique psychology cannot be explained by the stimuli of the external environment. The behaviorists tried it once and it simply did not work. Can we consider the semiotic system as a self-organizing system? Moreover, why should we do it in the first place? Can we not attribute the origin of signs to the Almighty or to the external environment that forces on our mind a given system of signs? Is it possible to consider signs as attractors? The idea of adopting the terminology of complexity science for inquiring into the sign system seems attractive since terms such as non-linearity, emergence, self-organization and attractor can easily be adopted for describing the dynamic of the sign system. However, this move should be done with caution. The next chapter explains why.

Chapter 17
Toward a Dialogical Complexity

Summary: Models and metaphors adopted from the natural and exact sciences seem to prevail in the study of the human. This chapter explains why the ubiquity of these models seems to miss the essential of the human, and at the same time why the terminology of complexity may be adapted for studying our sign system.

One of the most basic difficulties in understanding a given phenomenon consists of modeling it through an appropriate *form of representation*. I use the term "form of representation" and not the term "representation" per se, because the problem of using an appropriate representation for modeling a given phenomenon is a salient one. Less salient is the fact that a particular representation is a specific expression of different forms of representation that cannot easily be exchanged or applied to the same sorts of problems. Thus we cannot meaningfully represent a piece of music by using a mathematical model (or vice versa). Try to play Euclid's elements on a piano or to translate Pergolesi's Stabat Mater into mathematical equations in order to sense the meaning of representational orthogonality.[42]

The above experiment may be considered trivial in the sense that a reasonable person does not really believe that such a hypothetical translation of the two semiotic systems is possible or meaningful. However, similar difficulties in translation are evident, but less salient, in the humanities and social sciences where fashionable models are embraced from the exact and natural sciences without critically examining the meaning of this activity. A classic example of this is Kurt Lewin's attempt to apply topology to social psychology (1936). Jean Piaget (1968) wittily commented that he is not familiar with any psychological theorem that was proven based on this venture.[43] Today, we see similar attempts when researchers apply chaos, fractal geometry and other highly fashion mathematical tools to various social phenomena. While metaphors are always welcome, the meaning of those metaphors and their benefits should be closely examined.

At this point, I would like to explain why it is so difficult to move between the two cultures and to apply the notion of complexity to social phenomena. I would like to explain my objection to adopting a naive theory of complexity by disclosing the syntactic form of representation that governs the natural/exact sciences.

A syntactic form of representation describes the *rules* (or interactions) that govern the behavior of a set of signs (or tokens) abstracting both from their semantics (their meaning) and their pragmatics (the way they are actually used). A syntactic form of representation has several properties that differentiate it from other forms of representations. The first property is that a syntactic representation/model is characterized by *meaningless* tokens. We may define a token as "meaningless" when its contribution to the whole of which it is a part is totally determined by the rule/s (the syntax) that connect it with other tokens that compose the whole, and not by its semantics, which is always particular and unique. For example, consider the familiar logical syllogism:

1. IF "A" THEN "B"
2. "A"
3. THEREFORE "B"

Whatever the meaning of propositions A and B, the structure of the syllogism is always valid. Therefore, whether "A" means "the cat is white" or "OH! OH! Baby 69" is of no importance. The same is true concerning complex systems in which the agents are a well-defined set of components. Although the tokens in a syntactic model are meaningless in the sense previously used, they are meaningful in another sense. They are meaningful in the sense that they can easily be identified as the objects of our inquiry. In other words, since syntactic models are context independent (they lack a semantic or a pragmatic aspect), they hold some kind of an abstract Platonic status and their components seem to be self-evident.[44] This property clearly differentiates between the natural/exact sciences and the humanities/social sciences in which the objects (the components) of the system are continuously debated.[45]

Another important property of the syntactic form of representation is that it involves "horizontal" (or first order) rules for describing the behavior of a system, that is, rules determining the totality by operating on the same logical level of analysis. For example, the interactions between the units in connectionist models are described by various procedures of activation/inhibition from which patterns emerge. Although these patterns are phenomenologically distinct from the interactions that underlie them, they are "mechanistically" embedded in "lower order" interactions. Therefore, a syntactic form of representation is usually used for describing a bottom-up process of emerging phenomena out of local interactions among homogenous and meaningless tokens.

Due to the enormous success of the natural and exact sciences, the syntactic form of representation has achieved undisputed dominance. However, there are many situations, especially concerning living systems, in which meaning and pragmatics must be taken into account in order to understand the behavior of the system under inquiry.[46] One of those systems is natural language. For example, consider the utterance:

There is a lion behind you.

In order to understand the meaning of this utterance we must consider not only its syntax but also its semantic and pragmatic aspects. In order to understand the whole utterance we must understand the meaning of the components. However, how do we understand the meaning of the components? For example, what is the meaning of the sign "lion"? It is a token of the whole utterance, but its meaning cannot be derived from the other words in the utterance or from the syntactic rules that govern the structure of the utterance. On the other hand, we cannot dismiss the token "lion" as meaningless because its meaning determines the meaning of the whole utterance. Does a lion mean predator? Does it mean a hairy person? Therefore, the meaning of the sign "lion" is important in terms of concrete behavior (e.g., run or laugh), but it is totally dependent on the context of the utterance (a safari in Africa or a talk at the university), which cannot be grasped in syntactic terms per se.

The attempt to consider sign systems through the lens of complexity sciences seems to be highly relevant. However, anyone who is familiar with the history of the humanities and the social sciences is well aware of the shortcomings of metaphors adapted from the natural sciences to the poor cousin, the humanities. Sometimes it even looks as if the social sciences are motivated by a version of Penis Envy we may describe as "Natural Science Envy." Each achievement of the natural sciences is translated immediately into new theories in the human sciences that try to explore unknown territories of the mind through the new tools available from the mature, richer cousin. Metaphors can be illuminating. However, they might also be misleading. Previously, for example, we used the game of chess as an example of a complex system. However, social/human systems are much more complicated since, as reflective beings, people do not simply play according to the "rules of the game" – they sometimes change the rules! Recall the crazy games in which Alice participates during her adventures in Wonderland. The rules are invented during the game and it seems that they are even secondary to the pleasure of inventing them! What happens here? Does the human subject resist scholarly study? The answer I would like to give is that the human subject is indeed a unique creature. (S)he is not the same as the components of physical systems. (S)he is not an object among objects. As Bakhtin, following Dilthey, suggests, "the entire methodological apparatus of the mathematical and natural sciences is directed toward mastery over reified objects that do not reveal themselves in discourse and communicate nothing of themselves" (Todorov 1984, p. 15). In contrast, studying the human as a dialogical subject, as a system that constitutes its being through a socio-semiotic process, invites a different method of inquiry, what, following Dilthey, we may describe as hermeneutical inquiry. This method obliges us to confront the unique nature of the human. As Rickman presents it, "Some of the distinctive features of hermeneutics are the focus on individuals, *the hermeneutic circle, the absence of an absolute starting point, and the direct confrontation of complexities instead of treating them from the outset in terms of the supposed constituents*" (1988, p. 58, my emphasis). Complexity sciences provide us with a unique and powerful dictionary for dealing with this uniqueness, a dictionary that bridges the gap between the two cultures. This dictionary may be useful as long as it takes into account the unique human situation and its socio-somatic-semiotic character. In the next chapter, I present the idea of the sign as an attractor, develop the logic of distinction to show that the social is an inevitable mode of their existence, and point to the social interactions between subjects as constituting the order (and therefore the reification) of our phenomenal world and at the same time explain the arbitrariness of the sign as a chaotic social process.

Chapter 18
The Architectonics of the Mind

The entire aesthetic world as a whole is but a moment of Being-as-event, brought rightfully into communion with Being-as-event through an answerable consciousness – through an answerable deed by a participant.
Bakhtin

Summary: A sign is an indication, the name we give to a distinction. How is it possible to settle the tension between the dynamic and indeterminate nature of a distinction and the allegedly fixed nature of the sign? Between the sign as located in the systemic closure of the individual and the sign as a social process? The answer is that signs can be considered attractors, stable configurations that emerge out of distinctions and through communication between people. In this context, the reification of the world and the chaotic nature of the sign are an almost inevitable result of an internal dynamic that characterizes the sign system.

Let us push the idea of distinctions and boundary one step further by using the theory of Bakhtin and the terminology of complexity science. To recall, Saussure's semiotic theory suggests that signs are arbitrarily associated with that which they signify and that they receive their meaning only by holding a position within a closed system of signs. Signs receive their meaning only by being differentiated from one another. In this sense, the semiotic system is a network of indication. If we consider the sign as a name of a distinction, a boundary phenomenon, this position confronts us with a mystery. On the one hand, we learned that an indeterminate and vibrating process of oscillation is what constitutes a distinction. On the other hand, a sign looks like a fixed entity; a pipe is a pipe and a cigar is (sometimes) just a cigar. So, how should we resolve the tension between the dynamic that underlies a distinction and the stability that appears to us when we are dealing with a sign?

A possible explanation for this mystery is that *the indication of the distinction, i.e., the act of signification, which is social through and through, is responsible for fixing the paradoxical dynamics of the distinction into a stable attractor, a sign.* Therefore, our sign system may be considered to be a network of attractors, stable configurations that fix the indeterminate flux of being-in-the-world. This network of attractors is arranged in a dynamic and recursive-hierarchical structure, which allows for the attribution of meaning to a given sign through the super-ordinate structure of which this sign is a part and of which it is a unit. This interpretation is in line with Volosinov's semiotic theory, since it posits the signs as the unique atoms of our phenomenological experience.

If we accept the suggestion that signs are some kind of attractors, stable appearances of an underlying flux, then we should encounter two main tasks. The first is to describe the unique dynamic that transforms us from the concrete, operational, singular and dynamic form of existence with dynamic objects in our indeterminate horizon to the relatively stable and abstract symbolic form that characterizes the sign system of human beings. The second task is to try and describe a dynamic that leads from the basic and universal experience of the world as experienced by all human beings through their bodies into a fragmented world populated by arbitrary symbols, that is, to deal with the symbol-grounding problem. The latter issue returns us to the question: How is it possible to transcend our systemic closure as unique beings and to use symbols in order to communicate with others? In this context, Bakhtin (1981) calls our attention to the fact that even when we have the illusion of transcending our systemic closure and approaching objects in reality, experienced objects, in their full experience, are the objects of our systemic closure:

> And what I grasp is not the object as an externally complete image, but rather my tactile experience corresponding to the object, and my muscular feeling of the object's resistance, its heaviness, compactness, and so forth. (p. 43)

It is amazing that this idea was presented long ago, before complexity science became fashionable and terms such as "autopoiesis" became cultural keywords. From Bakhtin's illuminating statement, we may conclude that the primary source of our mind is a number of distinctions that correspond to our somatic coupling with the environment, not abstract representations that correspond to the thing-in-itself. In other words, since our contact with the environment is constructed by our unique boundary structure, our perceptions are, to a large extent, the most concrete expression of that structure. According to this interpretation, the mind is not a projection of the outside world, but the expression *of our boundary structure as manifested in different planes of abstraction and complexity.*

Let us dwell on this idea. The most basic form of existence is purely operational and primordial being-in-the-world. In this context, the most basic activity of *signification* is the activity that constitutes the differentiated units: self and non-self, the observer and his environment, but not in terms of an object that precedes mind, rather in terms of certain boundary dynamics that bring forth reality and mind. A process of signification which is the construction of a boundary. A more advanced semiotic phase of development appears when the differentiation process becomes refined, and the exterior environment (that is, basic boundary relations!) is experienced, according to the boundary structure of the being, as more and more differentiated. This is the phase in which a child discovers not only his differentiated self as opposed to the environment, but the differentiation of forms *within* the environment. This refinement of the differentiation process is evident when the infant learns to differentiate between his/her mother (or father) and other people, strangers. Although fear of strangers is a qualitatively different developmental phase, it is, in essence, the same differentiation process provided by the resources of memorization and representation in the sense of operation previously described. The same process also holds for the differentiation of the inner environment where the infant starts to differentiate his sensual modalities. From this process of fine differentiation inside and outside of the organism, the way to concept formation and symbolization is potentially open, first, by differentiating (through the self-referential dynamic of the primary distinction) the operation/representation that exists on the first order level of analysis from its differentiation on the second order level of analysis. By differentiating (reflecting) on the difference (boundary relations) between the system and its exterior environment (or perceptual portions), a second-order difference is created which still preserves the physical forces that map the exterior environment into the system through the mapping function (the boundary). Here, a concept is primarily a process embedded in the exterior environment and which preserves its perceptual characteristic. However, at the same time, and due to the fact that it exists as a differentiation of system-environment relation (second-order differentiation), it is also a differentiated phenomenon of the observer. This process clearly portrays the dynamic leading from the primary distinction, as a self-organizing system that operates at different levels, to the emerging symbol, which is a modus of this system in its expanding activity. Note that this process is explained from within by the same dynamic that constitutes the primary distinction. It is as if the dynamic of the primary distinction is pushed "outward" and expands to higher (deeper) levels of abstraction.

    The symbol is created by elaboration of the differentiation processes (conceptualization) at a higher level. It emerges from the same differentiation process, but this time at a third-order level of analysis, in which concepts (conceptualization processes) are differentiated from each other, stripped of their perceptual cloth and produce a form of representation that exists as meaningful (that is, differentiated) only by their difference from each other (a la Saussure).

    The process just portrayed emphasizes the unique dynamic of the primary distinction as a recursive and hierarchical process that pushes itself outward (expands) to higher levels of abstractions. This description, however, is deficient since the dynamic of the primary distinction may produce expansion on a single level of analysis, but does not seem able to explain the shift to higher levels of abstraction. The next sections aim to ponder this difficulty.

18.1. SECONDARY DISTINCTIONS AND THE TRANSGRADIENCE OF CONSCIOUSNESS

In "Principles of Biological Autonomy," Varela (1979) makes a distinction between the *organization* and the *structure* of a system:

The *relations* that define a machine as a unity, and determine the dynamics of interactions and transformation it may undergo as such a unity, we call the *organization* of the machine. The actual relations that hold between the components that integrate a concrete machine in a given space constitute its *structure*. (p. 9)

The primary distinction is the organization of all living systems in all their forms. It is what we previously described as the universal structure of the sign system. Following Volosinov, I would like to suggest that the *social, or more accurately the socio-somatic-semiotic,* is the unique structure that makes material this abstract organization in the realm of human beings.

In order to explain this idea, let us move to the next phase of our inquiry. We know that our world is populated not only by a primary distinction that constitutes the identity of each of us, but also by numerous secondary distinctions that distinguish our self/non-self. A secondary distinction involves a similar logic to the primary distinction with one exception: It seems that the boundary that constitutes the secondary distinctions is not an integral part of the observer and it does not constitute the observer, but rather her differentiated universe. This statement is false and it will be clarified in a minute. A minimal secondary distinction can be represented graphically as follows, where point A represents the primary distinction (the observer), point B a differentiated primary distinction in her universe (another observer), and the arrows the oscillation between the observer and another observer that constitutes the secondary distinction:

Figure 2. The secondary distinction

Let us first examine each of the primary distinctions (observers) in itself. Each of the above distinctions is constituted through the process previously described. Following Kempe (1886), we can describe the left distinction (A) and the right distinction (B) as units indistinguishable from one another in the sense that both of the observers have the general triadic structure of a distinction: inside, outside and boundary. Like a sign, the observer's identity is constituted only by differentiating itself from the other. However, we can see that although the units A and B are indistinguishable, the pairs AB and BA are distinguished from one another since the *direction of the oscillation from A to B and from B to A can never be the same*. As Kempe argued, "Pairs [and even higher order collection of units] may in some cases be distinguished even though the units composing them are not" (1886, p. 3).

Simple as it may seem at first glance, the asymmetry of AB and BA, that is, the asymmetry of observers as systemic closures, the asymmetry that results from the internal dynamic of each observer, has far-reaching theoretical implications. Looking at the above distinction from a third-person perspective (the perspective of the outsider), one can describe it as AB or BA. Regardless, each observer must constantly shift between the two perspectives in order to grasp the full meaning of this distinction, a distinction that exists on a higher level of analysis than AB and BA, and a distinction that constitutes its differentiated identity! A metadistinction, a frame, which is necessary for the sub-distinctions. This shift to a higher level of (social) abstraction brings complexity into the picture, since the existence of each agent becomes dependent on the existence of other agents in the system and the way they define its identity.

Thus, any activity of differentiation from the observer's point of view necessarily involves the existence of the other, and a shift to a higher level of the recursive hierarchy, a shift which is embedded in the internal dynamic of every living system.

Moreover, AB and BA are *distinctive* but also *complementary* aspects of the same system since only the oscillation between AB and BA and the dynamics that produces a higher order distinction constructs the "full" grasp of the whole system. This notion can be described by the Bakhtinian term *"transgradient"* that designates "elements of consciousness that are external to it, but nonetheless absolutely necessary for its completion, for its achievement of totalization" (Todorov 1984, p. 95). Let us delve into Bakhtin's architectonic conception of the mind in order to understand the idea that the other (our relations with the other) is necessary for our existence and for the symbolization process.

18.2. THE ARCHITECTONICS OF THE MIND AND THE REIFICATION OF THE UNIVERSE

Mikhail Bakhtin (1895-1975) is one of the most interesting intellectual figures in the twentieth century. Bakhtin's writings are usually familiar to those interested in literary criticism or Slavic studies. However, Bakhtin produced a fascinating corpus of texts that cannot be grasped in disciplinary terms per se, and its relevance to the general venture of systems research is enormous. Being a part of a small intellectual circle that included Volosinov, Bakhtin's contribution to the idea of systems theory (or holistic thinking) has not been realized yet. However, the similarity between the ideas presented in Bakhtin/Volosinov's writings and the ideas presented by major holistic thinkers such as Gregory Bateson is simply amazing.

I will use Holquist's introduction to the collection "Art and Answerability" in order to present Bakhtin's conception of *architectonics*. The concept of architectonics concerns the way in which things are assembled into a unified whole. More specifically, Bakhtin was bothered by the question how relations between subjects are organized into relations of "I" and "another." In a sense, this is the same problem that bothered us previously when we inquired into the phenomenological outcomes of the primary distinction: The observer ("I") and its environment (the "other"), the self and the non-self. While *architectonics* is the general study of "how entities relate to each other," Bakhtin presents the term *aesthetics* to mean "the problem of consumption, or how parts are shaped into wholes" (Holquist in Bakhtin, 1990, p. x).[47]

According to Bakhtin, a whole is always wholeness as long as it is conceived to be wholeness by a given observer. This is a source of both power and difficulty in Bakhtin's theory, the attempt to generalize the uniqueness of the subjective experience, to transcend the primordial experience and to transform it into a current coin. In fact, Bakhtin expresses the uniqueness of the subjective experience in what Holquist describes as "the first law of human perception," "whatever is perceived can be perceived only from a uniquely situated place in the overall structure of possible points of view" (p. xxiv). This position complements the idea presented above concerning the similarity of the observers. Indeed, we are all the same in the sense that we are all distinctions. However, at the same time we are all unique since our differentiation, the differentiation that constitutes our identity, provides us with a unique point of view. In a sense, this "general law of uniqueness" stresses the existence of the observer as a unique epistemic closure that at its most basic form of existence interacts with singularities in flux. Surprisingly, this position, which is implied from our definition of the observer and the primary distinction as a Holon, does not imply a Solipsist conclusion. On the contrary! We realize our uniqueness only through the existence of others. "We are all unique but never alone" (p. xxvi).

Bakhtin further elaborates two more important points concerning the striving for wholeness. First, wholeness is not given: It is achieved. The subject must put his efforts into constructing wholes out of the chaos that surrounds him. This idea provides a clear link to the conception of living systems as *enactive* systems. As suggested by Dupuy and Varela (1992, p. 20): "We propose the term enactive to emphasize the growing conviction that cognition is not the representation of a pre-given world, but is, rather, the enactment of bringing forth of a world on the basis of history and the variety of effective actions that a being can perform." This

The Architectonics of the Mind 149

theoretical position clearly follows Merleau-Ponty (1962) and the rejection of the representational approach in cognition. That is, the world is not given and re-presented, as argued by traditional empiricism; nor does it emerge from pre-given conceptions, as suggested by the rationalists. The world is evident in the organism and the organism is evident in the world through the organism's unique form of "being-in-the-world," through the totality of its relations with the environment and with others that populate this environment.

The second point discussed by Bakhtin is that the parts are always in a state of dynamic tension, in *activity;* they are always in process, the same as the units of the primary distinction. However, out of this dynamic stable structures emerge. In our terms: a*ttractors.* The observer is involved in "making sense out of the world by *fixing the flux of its disparate elements*" (p. xxiv, my emphasis), although Being is a flux:

> What in life, in cognition, and in performed actions we call a *determinate object acquires its determinateness, its own countenance, only through our relationship* to it: it is our relationship that determines an object and its structure, not conversely. (p. 5)

This is an interesting statement because is blurs the difference between ontology and epistemology. Objects become objects only through the *relationship* of certain observers to them and not through their own independent presence in an indeterminate horizon of Being. The observers, being the complementary part of the equation, exist as differentiated observers, as long as they have a certain relation to other observers. The objects are conceived as "independent" and therefore reified when the observer speaks about objects as detached from the relationships that constitute their existence. This point is especially important, since what it actually says is that *reification begins: when the dynamic that constitutes our phenomenal world is forgotten;* when we speak about objects (i.e., signs) as detached from the dynamics that give them birth; when we exchange the metaphysics of process in favor of a metaphysics that offers us a look at the universe as reified, and on its singularities as individuated objects, as self-consummated objects identified with themselves. The notion of identity is important since identity is a rhetorical term that allows us to speak about the object as if it has some kind of an independent existence.

Bakhtin's discussion concerns the artistic work and, therefore, he illustrates the above ideas by using the author of a novel and its hero:

> When, on the other hand, an artist undertakes to speak about his act of creation independently of and as a supplement to the work he has produced, he usually substitutes a new relationship. (p. 7)

This relationship is what constitutes the reified universe. The "it" in our world. A relationship that substitutes the co-occurrence of an observer and her environment, with a *meta-relation*, a reflection of this relationship. This meta-relation, reflection on the actual occurrence, through the use of language, is the unavoidable meta-frame, the context, that separates mind from the world and the observer from the object. It is the inevitable result of the dynamic that constitutes our sign system. What is the rule of the other in this dynamic? The other is necessary for us in order to achieve a "full" picture of ourselves from the outside and therefore for fixating the flux of being into normative stable appearances (attractors) we call signs. The others are the invisible frame of the sign. To recall, Laing's "Knots," we cannot see ourselves from the inside except when we move to the outside ... well, the price is evident. For example, the social aspect of Being allows us to see ourselves from the outside as complete "wholes," as demarcated and consummated wholes, as objects defined by others, but at the price of objectification. However, the other being, a partial and alien position to my own existence, is also a source of chaos while I am translating myself into his mind. While meaning is fixed by the interacting individuals as a necessary form for constituting their identity, the price is that the mapping from one agent to another at the macro-level of analysis, and for the long-run, necessarily leads to the differentiation of the sign from its conceptual cloth, as errors in translation are propagated throughout the system. In this sense, the other is both a source of order by fixating the flux of Being into stable structures, and also a source of dynamic that explains the way the deterministic logic of distinction results in arbitrary symbols. Let me explain this idea further.

We may interchangeably use the terms "author" and "hero," "observer" and "other" in order to adapt Bakhtin's ideas to our previous discussion. The observer is the whole (the primary distinction) that consumes (contains) the parts of its existence the same as the author consumes the hero and the artistic work. This statement simply points at the observer as an epistemological closure. There is nothing in mind beyond mind itself. This is an autopoietic system. This whole, however, is transgradient to the parts of which it consists, since it always consumes them from a higher level of abstraction: "The hero's consciousness, his feeling, and his desire of the world...are enclosed on all sides ... by the author's *consummating* consciousness of the hero and his world" (p. 13). Consuming the world means that the world exists as long as the observer consumes it, and, at the same time, being outside of the world as a differentiated whole provides the observer with a unique perspective of his own being. The observer uncovers himself through the existence of the world outside of himself and, more specifically, constitutes his existence through the "otherness" of the world that exists in his own mind. The importance of the other for the constitution of one's whole existence is expressed by Bakhtin in a beautiful way:

> When I contemplate a whole human being who is situated outside and over against me, *our concrete, actually experienced horizons do not coincide*. For at each given moment, regardless of the position and the proximity to

me of this other human being whom I am contemplating, I shall always see and know something that he, from his place outside and over against me, cannot see himself ... As we gaze at each other, two *different worlds are reflected in the pupils of our eyes.* (p. 23, my emphasis)

This position calls for a shift of the observer from his unique position in the universe as an undefined singularity to the position of the other:

I must empathize or project myself into this other human being, see his world axiologically from within him as *he sees* this world; I must put myself in his place through the excess of seeing which opens out from this, my own place outside him. I must enframe him, create a consummating environment for him out of this excess of my own seeing, knowing, desiring, and feeling. (p. 25, my emphasis)

But how can I see myself from the outside in order to constitute my identity as a differentiated whole? This problem of *"translating" myself from singularity into communicativity* is solved by objectifying my singularity through the normative medium of signs:

How to accomplish the task of translating myself from inner language into the language of outward expressedness and of weaving all of myself totally into the unitary plastic and pictorial fabric of life as a human being among other human beings, as a hero among other heroes [that is, as an individual]. (p. 31).

The question is, of course, how is it possible for the mundane observer, for this "close" wholeness, to project himself into the mind of the other in order to constitute her own existence? The answer is *through signs. Through symbols* that constitute what we may describe as thinking:

Thinking has no difficulty at all in placing *me* on one and the same plane with all *other* human beings, for in the act of thinking I first of all abstract myself from that unique place which I – as this unique human being – occupy in being; consequently, I *abstract myself* from the concretely intuited uniqueness of the world as well. (p. 31)

And:

... the actual horizon of my life, has a *twofold* character, because *I* and the *others* – we move on different planes of *seeing* and *evaluating* (not abstract, but actual, concrete evaluating), and, in order to transpose us to a single unified plane, I must take a stand axiologically *outside* my own life and perceive myself as an other among others. *This operation is easily accomplished by abstract thought* ... (p. 59, my emphasis)

That abstraction, in the sense of a symbolic process, takes me from singularity to communicativity and individuality. Merleau-Ponty also discussed this point:

> There is one particular cultural object, which is destined to play a crucial role in the perception of other people: Language. In the experience of dialogue, there is constituted between the person and myself a common ground; my thought and his are inter-woven into a single fabric, my words and those of my interlocutor are called forth by the state of the discussion, and they are inserted into a shared operation of which neither of us is the creator. We have here a dual being, where the other is for me no longer a mere bit of behavior I the transcendental field, nor I in his; we are collaborators for each other and we co-exist through a common world. (1962, p. 354)

Therefore, abstract thought – our semiotic system – our mind as a semiotic interface, is the ultimate source of experiencing myself from the outside, and the source of our reified universe. This should not necessarily be taken for granted because my position to the world in terms of what there is in the world can have a very different nature.

Bakhtin suggests that there are two possible ways of approaching the outside world: from within (as the horizon of the observer) and from the outside (as the environment of the observer). When I experience the object from within it "stands *over against me* as the object of my own ... directedness in living my life." As a primordial experience. In this context, the object is "a constituent of the unitary and unique *open* event of being" (p. 97). This is a world of actions not of given objects "From within my own consciousness – as a consciousness participating[48] in being – the world is the object of my acts of doing" (p. 98). However, when we turn to the object from the outside, the world turns out to be a world populated by static objects that may appear as randomly associated with their origins. Two scholars who follow Bakhtin provide another layer to our analysis by differentiating between *consciousness* and *thinking* in a way which is highly relevant to our discussion:

> *Consciousness*, in its initial definition ("knowing-with-others"), is the reproduction of our perception, in our mind, of things' coexistence. Here is one object, there is another object, each of them is *identical-with-itself* [my emphasis], and both are regarded as perpetually coexistent. Thus, in a child's consciousness, mother is self-same – irrespective of whether she is beside, or has left and then returned in a minute (in an hour, in two weeks ...). Mother is in my consciousness not only when she is satisfying my immediate needs. And I exist not only in the moments of my immediate needy relation to her... As distinct from consciousness, *thinking* is engaged with the possibility – not the actuality – of being ... Thinking always puts a thing, an event, another human being at the point of their possible nonbeing, and therefore thinks of them as occurring anew. (Alexandrov & Struchkov, 1993, p. 352)

If we ignore the specific terminology of consciousness and thinking, we may realize that again our basic dynamic pushes us toward the reification of the world, toward a childish metaphysics of reflecting on our world through the scheme "mother is here." This form of "intellectualism" has been criticized Merleau-Ponty saying that:

> Intellectualism cannot conceive any passage from the perspective to the thing itself, or from sign to significance otherwise than as an interpretation, an apperception a cognitive intention ... But this analysis distorts both the sign and the meaning; it separates out, by a process of objectification of both, the sense-content, which is already "pregnant" with a meaning ... it conceals the organic relationship between subject and world. (1962, p. 152)

In sum, the internal dynamic of the primary distinction encapsulates the need for the other in order to constitute its systemic closure: We are all unique but we are never alone. We need the other in order to establish our own selves. In this sense, my existence never precedes the other or the environment. This dynamic in which I establish my own identity as a demarcated whole through the existence of the other forces me to project myself to the other and to see myself from his point of view. This dynamic, in which I see myself from the outside through the other is possible only through signs-frames that establish a platform for our communication. Signs are the tools for this process of alienating my self from my primordial experience in order to constitute my identity as a whole differentiated being. The signs we use for achieving this aim (and their accompanying frames) are, therefore, both the result of an "internal" dynamic of differentiation and an "external" dynamic of corresponding to others. This dynamic of fixation is possible only if we take signs to be identical with themselves. This is the fantasy that necessarily accompanies the social aspect of semiotics. If signs do not obey the law of identity and identify with themselves, there is no point in using them as a tool for mediating the correspondence between others and me. According to this interpretation, the law of identity does not point at "real" identity of the sign with itself, but at a pragmatic necessity, as a first principle for constituting communication and, hence, my differentiated existence. It is the metaphysical fantasy all of us should adopt in order to participate as humans in the game of Being. As Spencer-Brown has already commented, it is one of those value judgements we impose on a ship in order to prevent it from sinking.

In order to clarify the point further we can examine some definitions of identity. From a purely logical point of view identity is denoted as "the relation each thing bears just to itself." Beyond the formal logical context, this definition raises more theoretical difficulties than it solves regarding the mysterious nature of this relation. A more convenient strategy is to think about a criterion of identity rather than about the definition of identity. Grayling (1997) suggests that:

> A criterion of identity for something is that criterion by means of which we can individuate it, specify which one it is, tell where it begins and another leaves off; in short, by means of which we can pick something out or tell that it is the same one again. (p. 30)

It should be noted that Grayling's suggestion does not provide us with a criterion of identity, but just with a description of what it means to have a criterion of identity. In this sense, it leaves us too with enormous quandaries concerning identity. However, Grayling's suggestion is important in the sense that it points at identity as a kind of *differentiation process*. I fully accept the interpretation of identity as a differentiation process. However, reflecting on this process we may also consider identity as a meta-postulate or as a first principle for any act of symbolic communication. *Symbolic communication assumes that the symbols exchanged during the communication are fixed enough to provide their exchange beyond a particular and perceptually embodied context of a particular operation.* Although the sign is dynamic, we must assume its identity with itself, its objectification, as a condition for communicativity. In this sense, our "semiotic" form of existence as creatures dealing and communicating with abstract signs (= symbols) is possible only by assuming a first principle of identity.

This interpretation explains why Bakhtin argues that "aesthetic intuition [primordial experience] is unable to apprehend the actual event-ness of the once-occurrent event, for its images or configuration are objectified, that is, with respect to their content, they are placed outside actual once-occurrent becoming" (1990, p. 3). In fact, what characterizes "theoretical thinking" is that it creates a "fundamental split between the content or sense of a given act/activity and the historical actuality of its being, the actual and once-occurrent experiencing of it" (p. 2). "very thought of mine, along with its content, is an act or deed that I perform – my own individually answerable act or deed" (p. 3). This is the "lived" experience, "but one can take its content/sense moment abstractly, i.e., a thought as a universally valid judgement" (p. 3), and in this case lived experience (individual historical moment) does not exist anymore. The demarcation between the two worlds is not a problem as long as the theoretical world does not attempt to overcome the lived experience. "But the world as the object of theoretical cognition seeks to pass itself off as the whole world, that is, not only as abstractly unitary Being, but also as concretely unique Being in its possible totality" (p. 8), and Bakhtin even locates this attempt in its historical context of *modernism* in which the dynamic of the sign and its accompanying frame turned into a metaphysics of alienation.

What about the chaotic nature of signs? Why do we call a cat "cat" and others call it "gato"? The dynamic of our sign systems is

The Architectonics of the Mind

deterministic. The symbol emerges from the primary distinction through a deterministic process of differentiation, through recursive-hierarchy. This dynamic is also a non-linear dynamic in the sense that the state of the system at T0 must be taken into account while we examine the state of our system at T1. This non-linearity is a derivative of the recursive and hierarchical nature of our sign system, and its social aspect. However, why is this system so sensitive to initial conditions? After all, our sign system seems to be anchored in our universal body and to reflect our universal experience. The answer should be sought in our unique architectonic of the mind. The whole process of signification is based on the fact that we are "all unique but never alone." We objectify our basic and singular experience through the other. Therefore, any kind of translation from the singularity of the primordial experience to the individuality of the sign system ("a pipe is a pipe") is prone to variations of translations, variations that echo through the semiotic net to create the divergence of our sign system.

The biblical story of the Tower of Babel and the accompanying interpretation (Midrash) may be an excellent interpretative framework for closing the discussion concerning the chaotic nature of our sign system. In Genesis 11:1 we meet the generations that follow Noah and the catastrophe that exterminated almost all the human race. We are told that this generation migrated from the east, found a plain in the land of Shinar and settled there. Breshit Rabba tells us that this movement is not a purely geographical one, but a movement from the ancient to the *modern* world. One of the Midrash interpretations locates the story of Babel and the divergence of languages in a context of transition to a modern era in which human beings have the abstract language and the technological facilities to build cities with giant towers such as the Tower of Babel, and by that to encounter God himself. Those people have one language, a universal medium of communication through which they can gather in Shinar from all places and work together to encounter the Almighty. Is this the language of abstraction that may bridge the communicative gap between an American scientist and a Russian scientist working together to break the secrets of the genome and to encounter Creation itself? Is not the venture of Babel the venture of modernism with its striving toward the abstract, the universal and a "new scientific dawn"?

The situation of "one language" changed when people rebelled against God: "Come, let us build us a city and a tower with its top in the heavens, that we may make us a name, lest we be scattered over all the earth." The people are afraid of being scattered. They are afraid of the divergence imposed on them by God's command. This is a generation that still feels insecure recalling the great flood that exterminated Noah's generation and decides to control its own destiny by technological means. God is shocked by this attempt, decides to interfere with men's dangerous act, and says to his angels: "Come, let us go down and baffle their language there so that they will no longer understand each other's language." "Therefore it is called Babel, for there the Lord made the language of all the earth babble." The result of this baffling is a lack of communication between the different nations: "cat" for the yawning creature in English-speaking countries, and "gato" for the same yawning creature in Spanish-speaking countries. From this lesson, we can learn that the chaotic nature of language is associated with modernism, abstraction and the anti-divergence movement of people. Indeed, modernism, as a high point in the alienation of man from his world and the celebration of the abstract as the real, is an issue that is intensively discussed.

We close this book not by promoting some kind of nostalgic return to the pre-modern, but by arguing again that our mind is a unique combination of reason and fantasy. Those two activities result from the same source: our ability to detach from being-in-the-world and to reflect on our unique form of being-in-the-world that includes our ability to reflect on our unique form of being-in-the-world... It is only when reason overcomes fantasy (or vice versa) that we believe that we may transcend the boundaries of the mind and bridge the gap between mind-reality and reality through the power of reason (Tower of Babel?) or fantasy. When reason overcomes fantasy, as propagated in the rhetoric of modernism, we try to overcome that which is the possibility of cognition and turn it into a cogitate object. The world is not an object. Only by weaving together reason and fantasy we can keep our ship afloat.

Cat-logue Which is an Epilogue:
Where a Blind Man Ends

Dr. N: Well?
Bamba: Well, what?
Dr. N: What do you think about the book?
Bamba: I like your epilogue since I have great empathy with your ethical teachings. However, I would like to close this book with my own epilogue.
Dr. N: I'm not sure that you are academically qualified for that. Do you have a Ph.D.? Have you published enough in refereed journals? Have you been quoted enough by your colleagues?
Bamba: Luckily enough, I have not been disqualified from encountering life. So, please let me close this book with a wonderful fable and its lesson.
Dr. N: Fine. Let's hear.
Bamba: In one of his most famous pedagogical reflections, Gregory Bateson questioned the boundaries of the mind by asking where a blind man ends:

> Suppose I am a blind man, and I use a stick. I go tap, tap, tap. Where do *I* start? Is my mental system bounded at the handle of the stick? Is it bounded by my skin? Does it start halfway up the stick? Does it start at the tip of the stick? But these are nonsense questions...The way to delineate the system is to draw the limiting line in such a way that you do not cut any of these pathways in ways which leave things inexplicable. If what you are trying to explain is a given piece of behavior, such as the locomotion of the blind man, then, for this purpose, you will need the street, the stick, the man; the street, the stick and so on, round and round. (Bateson, 1973, p. 434)

Dr. N: Wonderful! What can we learn from this fable?
Bamba: First of all, it is amazing that Merleau-Ponty also uses the example of a blind man for arguing the same thing! Listen, he says that: "The blind man's stick has ceased to be an object for him, and is no longer perceived for itself; its point has become an area of sensitivity, extending the scope and active radius of touch, and providing a parallel to sight" (1962, p. 143).
Dr. N: I still don't get it. It is a wonderful, poetic fable, but ...
Bamba: Can't you see! The boundaries of the mind are indeterminate! Our mind as a semiotic interface can be constituted and reconstituted by the semiotic forms that form its activity to include technological artifactuals such as the blind man's stick. Metaphorically speaking, all of us are blind and all of us use different sticks to sense our being-in-the-world. When you suggested that a possible and positive outcome of this book is the

"reenchantment of the mind" you must have thought about this idea, about the possibility of pointing to our limitations, to our ignorance of the dynamic of life, to the frames through which we "legitimize" our systems of value and illusions, to those metaphysics that pretend to be physics… Please remember that if you consider mind as interaction then differentiation precedes identity and not the other way around.

Dr. N: Excellent summary, my dear cat. Indeed, we are all blind and we all find our way in the world through reason, through different semiotic sticks and fantasy. While dynamic objects provide the possibility and the constraint of reason, it is only fantasy which keeps our semiotic ship afloat. It is also a liberating idea that although reality constrains our cognition, fantasy can bring forth the myriad possible worlds that may expand from reality. Since your insights are so impressive, would you like to join the university and continue your scholarly research under my supervision?

Bamba: Definitely not. I just noticed a fat mouse running under the fence and I would like to inquire into its taste …

Dr. N: Sounds reasonable.

References

Aitchison, J. (1998). On discontinuing the continuity-discontinuity debate. In J. R. Hurford, M. Studdert-Kennedy & C. Knight (eds.), *Approaches to the Evolution of Language*. Cambridge: Cambridge University Press, pp. 17-30.

Alexandrov, D., & Struchkov, A. (1993). Bakhtin's legacy and the history of science and culture: An interview with Anatolii Akhutin and Vladimir Bibler. *Configurations, 2*, 335-386.

Anderson, J. R. (1993). *Rules of the mind*. Hillsdale, New Jersey: Erlbaum.

Anton, H., & Rorres, C. (2000). *Elementary linear algebra*. New York: Wiley.

Atlan, H. (1998). Immanent causality: A Spinozist viewpoint on evolution and theory of action. In G. Van de Vijver et al. (eds.). *Evolutionary systems*. Netherlands: Kluwer, pp. 215-231.

Bakhtin, M. (1973). *Problems of Dostoevsky's poetics*. USA: Ardis.

Bakhtin, M. (1981). *The dialogic imagination: Four essays by M. M. Bakhtin* (Trans. C. Emerson & M. Holquist). Austin: University of Texas Press.

Bakhtin, M. (1990). *Art and answerability* (Trans. Vadim Liapunov). Austin: University of Texas Press.

Bakhtin, M. (1999). *Toward a philosophy of the act* (Trans. Vadim Liapunov). Austin: University of Texas Press.

Barth, J. (1988). *The floating opera and the end of the road*. New York: Anchor Books.

Bavli Tractate Baba Mesia. (1996). A. Ch. 1 through 6. The Talmud of Babylonia (Trans. Jacob Neusner). Atlanta, Georgia: Scholars Press.

Bateson, G. (1973). *Steps to an ecology of mind*. London: Granada Publishing.

Bateson, G. (2000). *Steps to an ecology of mind*. Chicago: The University of Chicago Press.

Bateson, G., & Bateson, M. C. (1988). *Angels fear: Toward an epistemology of the sacred*. USA: Bantam Books.

Bateson, M. C. (1991). *Our own metaphor*. Washington, DC: Smithsonian Institution Press.

Berman, M. (1981). *The re-enchantment of the world*. New York: Bantam.

Brandon, R. N. (1996). *Concepts and methods in evolutionary biology*. Cambridge: Cambridge University Press.

Cheetham, M. A. (1991). *The rhetoric of purity: Essentialist theory and the advent of abstract painting*. Cambridge: Cambridge University Press.

Choi, S., & Bowerman, M. (1991). Learning to express motion events in English and Korean: The influence of language-specific lexicalization patterns. *Cognition, 41*, 83-121.

Chomsky, N. (1988). *Language and problems of knowledge*. Cambridge, Mass.: The MIT Press.

Chown, J. F. (1994). *A history of money: From AD 800*. London: Routledge.

Cohen, I. R. (2000). *Tending Adam's garden: Evolving the cognitive immune self*. New York: Academic Press.

Cohen, I. R. (2002). *Talmudic texts for visiting scientists: On the ideology and hermeneutics of science*. Manuscript in preparation.

Cole, M. (1996). *Cultural psychology: A once and future discipline*. Cambridge, Mass.: Harvard University Press.

Crump, T. (1981). *The phenomenon of money*. London: Routledge & Kegan Paul.

Deleuze, G. (1990). *The logic of sense* (Trans. Mark Lester). New York: Columbia UP.

Derrida, J. (1987). *The truth in painting*. Chicago: Chicago University Press.

Dilthey, W. (1923/1988). *Introduction to the human sciences* (Trans. Ramond J. Betanzos). Detroit: Wayne State University Press.

Duro, P. (1996) (ed.), *The rhetoric of the frame*. Cambridge: Cambridge University Press.

References

Dupuy, J. P., & Varela, F. (1992). Understanding origins: An Introduction. In F. Varela and J. P. Dupuy (eds.), *Understanding origins*. Dordrecht: Kluwer Academic Publishers, pp. 1-27.

Emerson, R. W. (1940). *The complete essays and other writings of Ralph Waldo Emerson*. New York: Random House.

Engelberg, J. (2001). *On the two-dimensional nature of human organism*. Unpublished manuscript. Office of Integrative Studies, University of Kentucky College of Medicine.

Francois, C. (1997). *International encyclopedia of systems and cybernetics*. Munchen: K.G. Saur.

Gablik, S. (1982). *Magritte*. London: Thames and Hudson.

Genesis Rabba: The Judaic Commentary to the Book of Genesis (1985). (Trans. Jacov Neusner). Atlanta, Georgia: Scholarly Press.

Galbraith, J. K. (1975). *Money: when it came, where it went*. UK: Penguin Books.

Glanville, R. (1990). The self and the other: The purpose of distinction. In R. Trappl (ed.) *Cybernetics and Systems '90*. Singapore: World Scientific, pp. 1-8.

Goux, J. J. (1990). *Symbolic economies*. New York: Cornell University Press

Grayling, A. C. (1997). *An introduction to philosophical logic*. London: Routledge.

Griffin, D. (1981). *Question of animal awareness*. USA: Rockefeller University Press.

Harnard, S. (1990). The symbol grounding problem. *Physica D*, *42*, 335-346.

Harre, R. A., & Gillett, G. (1995). *The discursive mind*. London: Sage.

Harries-Jones, P. (1995). *A recursive vision: Ecological understanding and Gregory Bateson*, Canada: University of Toronto Press.

Harrington, A. (1996). *Reenchanted science*. Princeton, New Jersey: Princeton UP.

Harris, R. (1987). *Reading Saussure*. London: Duckworth.

Harrison, A. (1987). Dimensions of meaning. In A. Harrison (ed.), *Philosophy and the visual art*. The Netherlands: D. Reidel Publishing Company, pp. 51-76.

Hegel. (1886/1993). *Introductory lectures on aesthetics* (Trans. Bernard Bosanquet). London: Penguin.

Herbst, D. P. (1993). What happens when we make a distinction: An elementary introduction to co-genetic logic. *Cybernetics & Human Knowing*, *2*, 29-38.

Hockett, C. F. (1960). Logical consideration in the study of animal communication. In W. Lanyon and W. Travolga (eds.), *Animal sounds and communication*, pp. 392-430.

Holquist, M. (1990). *Dialogism*. London: Routledge

Hookway, C. (1985). *Peirce*. London: Routledge.

Husserl, E. (1973). *Cartesian mediations*. Hague: Nijhoff.

Hussey, E. (1999). Heraclitus. In A. A. Long (ed.), *The Cambridge Companion to Early Greek Philosophy*. Cambridge: Cambridge University Press, pp. 88-112.

Jenny, H. (1974). *Kymatik*. Basel: Bislius Presse.

Johnson, M. (1987). *The body in the mind: The bodily basis of meaning, imagination, and reason*. Chicago: Chicago University Press.

Kahn, C. H. (1981). *The art and thought of Heraclitus*. Cambridge: Cambridge University Press.

Kauffman, L., & Varela, F. (1980). Form dynamics. *Journal of Social and Biological Structures*, *3*, 171-206.

Keller, H. (1928). *The story of my life*. Boston: Houghton Mifflin Company.

Kempe, B. (1886). A memoir on the theory of mathematical form. *Philosophical Transactions of the Royal Society of London*, *177*, 1–70.

Kilgour, M. (1990). *From Communion to cannibalism: An anatomy of metaphors of incorporation*. Princeton, New Jersey: Princeton University Press.

Kosko, B. (1993). *Fuzzy thinking*. New York: Hyperion.

Laing, R. D. (1970). *Knots*. USA: Vintage Books.

Lakoff, G., & Johnson, M. (1980). *Metaphors we live by*. Chicago: Chicago University Press.

Lakoff, G., & Johnson, M. (1999). *Philosophy in the flesh: The embodied mind and its challenge to western thought*. USA: Basic Books.

References

Levinas, E. (2001). *Nine Talmudic readings*. Jerusalem: Schocken (In Hebrew). [The original French version is: Neuf lectures Talmudiques, 1968, Les Editions de Minuit].
Lewin, K. (1936). *Principles of topological psychology*. New York: McGraw Hill.
Lewis, C. S. (1960). *Studies in words*. Cambridge: Cambridge University Press.
Luria, A. R. (1976). *Cognitive development: Its cultural and social foundations*. Cambridge, Mass.: Harvard UP.
Luria. A . R., & Vygotsky, L. A. (1992). *Ape, primitive man, and child: Essays in the history of behavior*. Orlando, Fla.: Paul. M. Deutsch Press.
Maturana, H. R., & Varela, F. J. (1972). *Autopoiesis and cognition: The realization of the living*. London: D. Reidel Publishing Company.
Maturana, H., Mpodozis, J., & Letelier, J. C. (1995). Brain, language and the origin of human mental functions. *Biological Research, 28*, 15-26.
Merleau-Ponty, M. (1942/1965). *The structure of behavior* (Trans. Alden L. Fisher). London: Methuen.
Merleau-Ponty, M. (1962). *Phenomenology of perception*. London: Routledge & Kegan Paul, Ltd.
McCulloch, W. (1970). *Embodiments of mind*. Cambridge, Mass.: MIT Press.
Miller, J. (1983). *States of mind*. New York: Pantheon.
Nash, W. (1992). *Rhetoric: The wit of persuasion*. Oxford: Basil Blackwell.
Neuman, Y. (1999). A difference that makes a difference: Reflections on artifact-mediated-consciousness. *Cybernetics and Human Knowing, 6*, 57-64.
Neuman, Y. (in press). Co-generic logic as a theoretical framework for the analysis of communications in living systems. *Semiotica*.
Neuman, Y., & Bekerman, Z. (2000). Where a blind man ends: Five comments on context, artifacts and the boundaries of the mind. *Systems Research and Behavioral Science, 17, 315-319*.
Newell, A. (1990). *Unified theories of cognition*. Cambridge, Mass.: Harvard University Press.
Nowak, M. A., & Krakauer, D. C. (1999). The evolution of language. *Proc. Nat. Acad. Sci. 90*, 8028-8033.
Peirce, C. S. (1978). Further consequences of the four incapacities. In *The Collected Papers of Charles Sanders Peirce*. C. Hartshore & P. Weiss (Eds.), Vol 5 and 6, pp. 156-190.
Piaget J. (1968). *Structuralism*. New York: Harper and Row.
Piattelli-Palmarini, M. (1980). (ed.). *Language and learning: The debate between Jean Piaget and Noam Chomsky*. Cambridge, Mass.: Harvard University Press.
Pinker, S., & Bloom, P. (1990). Natural language and natural selection. *Behavioral and Brain Sciences, 13*, 707-784.
Pinker, S. (1994). *The language instinct*. London: Penguin.
Plato. (1970). The dialogues of Plato. Translated by Benjamin Jowett. Vol 4. The Republic. London: Sphere Books Limited.
Poincare, H. (1952). *Science and method*. New York: Dover Publications, Inc.
Priest, S. (1998). *Merleau-Ponty*. London: Routledge.
Rajchman, J. (2000). *The Deleuze connections*. Cambridge, Mass.: The MIT Press.
Rickman, H. P. (1988). *Dilthey today: A critical appraisal of the contemporary relevance of his work*. New York: Greenwood Press.
Robertson, R. (1999). Some-thing from no-thing: G. Spencer-Brown's laws of form. *Cybernetics and Human Knowing, 6*, p. 43-57.
Sampson, E. E. (1996). Establishing embodiment in psychology. *Theory and Psychology, 6*, 601-624.
Sampson, G. (1980). *Schools of linguistics*. Stanford: Stanford UP.
Saunders, P. T. (1994). Evolution without natural selection. *Journal of Theoretical Biology, 166*, 365-373.
de Saussure, F. (1959). *Course in general linguistics*. New York: McGraw-Hill
Sebeok, T. A., & Danesi, M. (2000). *The forms of meaning*. Berlin: Mouton De Gruyter.
Simon, H. A. (1969). *The science of the artificial*. Cambridge, Mass.: The MIT Press.

Sober, E. (1993). *Philosophy of biology*. Oxford: Oxford University Press.
Spencer-Brown, G. (1979). *Laws of form*. New York: E.P. Dutton.
Suchman, L. A. (1987). *Plans and situated actions: The problem of human machine communication.* Cambridge: Cambridge University Press.
Thagard, P. (1996). *Mind*. Cambridge, Mass.: The MIT Press.
The Cambridge Dictionary of Philosophy (ed. R. Audi) (1995) Cambridge: Cambridge University Press.
Thom, R. (1983). *Mathematical models of morphogenesis*. New York: John Wiley.
Todorov, T. (1984). *Mikhail Bakhtin: The dialogical principle.* Minnesota: University of Minnesota Press.
Tomasello, M., & Akhtar, N. (1995). Two-years-olds use pragmatic cues to differentiate references to objects and actions. *Cognitive Development*, *10*, 201-224.
Turing, A. M. (1950). Computing machinery and intelligence. *Mind,* 59, 433-460.
Turner, B. S. (1996). *The body and society.* London: Sage.
Varela, F. (1979). *Principles of biological autonomy*. New York: Elsevier North-Holland.
Varela, F. (1992). Whence perceptual meaning? A cartography of current ideas. In F. Varela & J. P. Dupuy (eds.). *Understanding origins.* Kluwer Academic Publishers: Dordrecht, pp. 235-265.
Varela, F, Thompson, E., & Rosch, E. (1993). *The embodied mind: Cognitive science and human experience.* Cambridge, Mass.: The MIT Press.
Volosinov, V. N. (1986), *Marxism and the philosophy of language*. Cambridge, Mass.: Harvard University Press.
von Humboldt, W. (1992). The nature and conformation of language. In K. Mueller-Vollmer (Ed.). *The hermeneutic reader*. New York: Continuum.
von Foerster, H. (1974). *The cybernetics of cybernetics*, Urbana, Ill.: University of Illinois Press.
von Uexkull, J (2001). The new concept of Umwelt: A link between science and humanities. *Semiotica, 134*, 111-123.
Vygotsky, L. S. (1993). Introduction: The fundamental problems of defectology. In *The Collected Works of L.S. Vygotsky*, Vol. 2. The fundamentals of defectology. New York: Plenum Press, pp. 29-51.
Wells, H. G. (1947). *The country of the blind and other stories.* London: Longman Green and Co. Ltd.
Wittgenstein, L. (1968). *Philosophical investigations* (Trans. G. E. M. Anscombe). Oxford: Basil Blackwell.

Endnotes

[1] The idea of experimenting with ideas clearly adheres to Gillet Deleuze. See Rajchman (2000).

[2] It is unnecessary to say that intuition for Spinoza and Bergson was not identified with some kind of an animalistic way of thinking. For them, intuition is a higher phase in human development, a phase that has been described by transpersonal psychologists, such as Ken Wilber, as the transpersonal phase.

[3] Oscar Wilde said once, "It is style that makes us believe in a thing - nothing but style." Worth remembering while reading a "scientific" thesis.

[4] Nice metaphor. However, it seems to reflect not only Kant's dissatisfaction with skepticism, but also the general European dissatisfaction with a nomadic way of life. This rejection may find its roots in the biblical story of Cain, whose punishment was to wander around like a nomad. There is another position that emphasizes the importance of "intellectual nomads." For example, Benoit Mendelbrot, the father of Fractal Geometry, says that: "The rare scholars who are nomads-by-choice are essential to the intellectual welfare of the settled disciplines."

[5] Recall John Searle, who argued that a domain that describes itself as "science" is definitely *not* a science. For example, cognitive science, Christian science, and the behavioral sciences.

[6] This is of course a phenomenological study a la Husserl that does not seem to tell us a lot about reality in itself but about the way our mind reflects upon reality. I will get to this point later in the book.

[7] In general one may consider an object as a demarcated and relative piece of the universe, the subject of a physical or mental activity. The stability of this piece is depicted in the law of identity (A = A), stating that the thing always equals itself.

[8] Throughout the book, I usually use the term "phenomenon" as synonymous with "appearance."

[9] In this book I use the term "socio" as a tag for a collective form of activity and not in the sense of a sociological mechanism.

[10] In this context, it must be noted, the term "process" has been extensively used in many different senses and contexts. Therefore, it has become almost totally meaningless. However, a crucial majority of those who used the term "process" assume the existence of objects as preceding, both ontologically and epistemologically, the existence of certain processes in which these objects are involved. My opening theoretical stance is opposite in the sense that processes precede the existence of objects both from ontological and epistemological points of view. In this sense I believe that the way the idea of a process (Is it "a" process? Is it "it"? Is process an object?) is elaborated in this book is unique.

[11] This title is a paraphrase on Morris Berman's book: The re-enchantment of the world.

[12] In this book I use the term "behavior" although I do not adhere to its behaviorist sense. Following Merleau-Ponty (1942/1965) I consider behavior as neutral with respect to the classical distinctions between the "mental" and the "physiological" (p. 4).

[13] It is interesting that some researchers have argued that nouns are cognitively easier to handle than verbs (Tomasello & Akhtar, 1995). This finding may explain the reified universe. However, other findings point at the rule of cultural factors in the reification of the world. For example, verbal frequency in first words of a child have shown to be culturally dependent (Choi & Bowerman, 1991).

[14] For an anti-representational approach to the mind that follows the phenomenological tradition see Dreyfus, H. L. Intelligence without representation. http://www.hfac.uh.edu/cogsci/dreyfus.html.

[15] Although these approaches are more complicated and less differentiated than can be conveyed in theoretical discussions, I will present them through their ideal types in order to establish them as terms of reference for asserting my own argument.

[16] A kind of "direct perception" a la Gibson.

[17] Semiotics as defined by Saussure is 'the science of the life of signs in society."

[18] This idea also appears in the writings of Vygotsky, Volosinov and Bakhtin.

[19] The Darwinian evolutionary perspective is just one possible developmental approach to semiotics. However, due to its centrality I focus on it.

[20] "Hard work" in the simplistic physical sense of the term. After all, using words cannot be considered "easy work."

[21] Easily supported, of course, when you posit it against the creationist perspective with its limited explanatory power. There are some serious scientific challenges (and alternatives), though, to the Darwinian evolutionary theory. See, for example, Saunders (1994).

[22] This is a very important statement since it locates Dilthey together with Spinoza and Goethe as one of the fathers of holistic thinking and general systems theory.

[23] This statement seems strange to those who are familiar with the work of Vygotsky. Indeed, Vygotsky is the one who is most associated with the idea of thinking through language. This idea clearly represents Vygotsky's theory. However, as Cole (1996) suggests, it seems that Vygotsky misjudged the relation between language and thinking by placing the phylogenetic influence ahead of cultural ones without acknowledging the way that cultural history and phylogeny interpenetrated one another. This is the reason why the relation between language and thought is presented in his writing in what may be considered as a "double voice."

[24] The relation between language and thought has been dismissed by the wonder kid of MIT - Steven Pinker - in his celebrated book, "The Language Instinct" (Pinker, 1994). However, while discussing the relation between language and thought, Pinker refers only to the simplified, and theoretically vulnerable, version of Sapir-Whorf hypothesis and does not mention the socio-cultural point of view as propagated by Luria and Vygotsky at all. In this sense, Pinker is throwing heavy punches at a straw man without confronting a serious challenge to the theory he presents. Vygotsky and Luria's socio-cultural conception of the mind may be contested, but cannot be ignored. Therefore, I suspect that Pinker's disregard for Luria and Vygotsky reflects a wider problem of the academic milieu - theoretical sectoriality in which one reads and quotes only members of a specific academic milieu of which one is a part.

[25] This is not to be confused with the work of Bakhtin that has been intensively quoted and elaborated. A short survey in academic data bases may prove this point by showing the proportion of references to Volosinov's work in contrast with Bakhtin.

[26] A similar notion appears in the writings of Merleau-Ponty.

[27] Even the concept of "nature" has been transformed from an activity of giving birth in the Greek language into a static entity in modern English! Nature is one of the signs that have been interestingly transformed from a process to an object. The common meaning of nature refers to character or kind. We may speak of *humana natura*, the character of all men beyond individuals and generations. The Latin term for nature is *Natura* and it shares a common base with *Nasci* which means "to be born." Of special interest is the relation between nature and the Greek Phusis (beginning, coming-to-be). The meaning of *Phusis* is a sort of thing passed from *Phusis* to *Natura* and from *Natura* to Kind (Lewis, 1960).

[28] The semiotic property of money is evident in the common practice of throwing coins into fountains and making a wish. Since money (coins) has a semiotic value, it is a carrier of the wish, it is a medium through which the wish is transferred.

[29] There is another interpretation of abstractness as a technique, "a way to get behind appearances to the fundamental reality" (Cheetham, 1991, p. 11).

[30] Examining the socio-cultural development of the sign is beyond the scope of this book. However, it seems evident that modernism was the period in which the reification of the world has reached its peak through the sign-context fantasy of detachment from the concrete.

[31] This is the reason why abstract art has so much involved the study of geometrical figures (e.g., Mondrian) that do not exist in the concrete world.

Endnotes

[32] It appears in Baveli, Baba Metzia (pp. 286-7).

[33] This state of mind is clearly evident in Levinas's (2001) modern reading of the Talmud. For a very similar argument, see Cohen (2002).

[34] This conception points to the alleged obsession of God with man's behavior, an idea which is deeply rooted in the Kabbalistic and the Hasidic tradition of "Tikun Olam" ("betterment of the world" in English).

[35] The ourobouros can be traced to Greek where it was used as a symbol of time as cyclic. Christians adopted this symbol of the limited confines of this world (that there is an outside demarcation of an inside). In our case, the ourobouros signifies the unavoidable circularity that underlies every form of semiotic activity.

[36] See Merleau-Ponty: "In the action of the hand while it is raised toward an object is contained a reference to the object, not as an object represented, but as that highly specific thing toward which we project ourselves, near which we are, in anticipation, and which we hunt. Consciousness is being toward the thing through the intermediary of the body" (1962, pp. 138-9).

[37] At this point, it is important to mention Spencer-Brown's response to a question addressed by Gregory Bateson. Bateson was asking about the "then" of logic (i.e., If A THEN B) and if this "then" is devoid of time. Spencer-Brown's answer to this question is that since logic is devoid of self-reference, it is devoid of time. This is an important statement because if we would like to understand the behavior of living systems that unfold in time, classical logic is not the appropriate tool.

[38] The transcription of this conference, held in 1973, appears in: http://www.rgshoup.com/lof/aum/sssion1.html

[39] It is better to describe this process as if the observer and its environment were coupled in a periodic form.

[40] I would like to thank Peter Harries-Jones for introducing to me this concept.

[41] The only reference I am familiar with discussing Spinoza in the context of self-organizing system is Atlan (1998).

[42] This experiment raises some serious questions concerning Mentalese - the general language of thought. If such a language exists, why is it so difficult to translate different forms of representations? Is it the same difficulty evident when trying to translate from one language to another?

[43] In fact, Lewin was fully aware of the difficulties in applying topology to psychology. In this sense, his courageous intellectual attempt is presented and criticized in my book for tactical reasons only.

[44] This is definitely not the theoretical position I hold. Therefore, I emphasize that they "seem" to be incontrovertible and not that they are really so.

[45] The polemical nature of signs in the humanities may be considered as cognitively economical. "Meaningless signs" is the idea that the sign is arbitrarily associated with its signification. Although this idea may be contested, we cannot deny its cognitive economy; if a sign is arbitrarily associated with its signification, then the same sign may signify different things! In different contexts, the sign would have different meanings. We will get to this point later.

[46] Charles Morris has already mentioned this point concerning natural language.

[47] It must be noted that in contrast with mereology which is a system of logic that tries to formalize part-whole relations as they are, Bakhtin's venture always takes into account the rule of the observer and therefore his theory provides us with a source of inspiration while inquiring into the logic of the semiotic system.

[48] See Berman (1981).

Name Index

Bakhtin, M., 47, 97–99, 110–113, 141, 143–144, 147–150, 152, 154
Bateson, G., 1, 7, 12, 67, 76, 100, 102, 107, 111, 120–121, 148, 157

Deleuze, G., 1, 106, 133
Descartes, R., 5, 16, 18–19, 81, 99, 109, 128, 131, 133
Dilthey, W., 10, 34–35, 141, 159

Engelberg, J., 103

Goux, J. J., 57–58

Heraclitus, 71–73, 75
Harries-Jones, P., 102
Hegel, 38, 57–58, 61, 64, 77, 99, 128
Herbst, D. P., 99–100
Holquist, M.,107, 112, 148
Husserl, E., 17–20, 24, 59, 73, 109

Keller, H., 116–118

Lakoff, G., 22, 115–122
Luria, A. R., 43–44, 48, 59–60, 109, 122

Magritte, R., 4, 61–62, 64
Maturana, H., 104, 115

Merleu-Ponty, M., 1, 7, 11, 17, 19, 23, 55, 81, 83, 99, 104, 108, 110, 112, 115, 118–120, 122, 149, 152–153, 157

Neuman, Y., 48, 112, 121–122

Peirce, C. S., 7, 9, 20–21, 23–25, 32, 44, 49, 57, 81, 99, 104, 111–112, 132
Piaget, J., 6, 58, 122–123, 139
Plato, 42, 71, 75, 93, 105
Poincaré, H., 122

Saussure, F. de, 22, 29, 39, 41–47, 49–50, 52–53, 84, 91, 143, 145
Spencer-Brown, G., 4, 89–91, 93, 95–96, 103, 107, 113, 132, 153

Turing, A. M., 115
Thom, R., 2–3

Varela, F., 91–95, 115, 146, 148
Volosinov, V. N., 7, 39, 44, 46–54, 65, 68, 76, 111, 144, 146, 148
von Uexkull, J., 90
Vygotsky, L. S., 10, 43–44, 59–60, 116–117

Wells, H. G., 116–118
Wittgenstein, L., 15, 37, 43–44, 59

Subject Index

Abstraction, 11, 13, 50, 57–59, 67, 107, 111, 144–145, 147, 150, 152, 155–156
Activity and Vygotsky, 10
Appearances, 7, 9, 17–19, 71–73, 75–76, 81, 83–84, 89–91, 93–94, 103, 144
Architectonics, 143, 147–148
Artifactuals, 121–122
Attractor, 128, 135–138, 142–144, 149–150
Autopoiesis, 144

Being
    Being-in-the-World, 1, 9, 11, 13–14, 55, 70, 75, 110, 115, 119, 143–144, 149, 156–157
    Flux of Being, 9–11, 70, 88, 112–113, 143, 150
    Horizon of Being, 149
Blind, 117–118, 157–158
Body, 35, 38, 102–103, 108, 113, 115, 117–123, 155
Boundary, 11, 48, 52, 85, 90, 93–95, 97, 99–100, 102–105, 107, 110, 113, 120–123, 125, 143–146

Chaos, 135–136, 148, 150
Circularity, 38, 68, 76, 81–82, 100, 104
    Hermeneutic circle, 76, 100–102, 141
Closure systemic, 93, 102–104, 107, 125, 138, 143–144, 148, 150, 153
Cognition, 4, 10, 17, 20–21, 31, 44, 49, 51, 55, 101, 108, 148–149, 154, 156, 158
    Cognitive science, 93, 103, 115

Complexity, 48, 57, 61, 123, 125–131, 135–136, 138–139, 141–144, 147
Concept, 11–13, 20, 22, 42–43, 45, 57–59, 145
Consciousness, 6, 8, 15–19, 22, 23, 25, 33, 49–50, 80–81, 119, 122, 143, 146, 147, 150, 152–153, 161
Context, 13, 22, 31, 45, 51–53, 56, 59, 61, 63, 65, 69, 76, 101, 110, 121, 125, 140, 141, 150, 154
Culture, 6, 80, 86, 99, 105, 118–119, 121

Differentiation, 11, 14, 19, 63, 82–84, 89–93, 107–108, 144–148, 154–155, 158
Distinction
    Primary, 90–97, 99–100, 102, 104, 107–109, 111, 125, 127, 131, 145, 146, 148–150, 153, 155
Dynamic object, 9–10, 12, 20, 68, 70, 108, 110, 144, 158

Embodiment, 2, 58, 70, 118–123
Emergence, 127, 135, 138
Enactive, 148
Epistemology, 4, 20, 58, 72, 75, 79, 80, 83, 109, 122, 131, 149
Evolution
    Cultural, 55, 58
    Darwinian, 4, 20, 58, 72, 75, 79, 80, 83, 109, 122, 131, 149

Fantasy, 79, 153, 156, 158
Frame, 61, 64–65, 67–70, 76, 110, 150, 154

Subject Index

Holistic, 35, 53, 55, 83, 100, 119, 128, 132, 148
Holon, 76–77, 100, 104, 148
Identity
    of the sign, 10, 13, 45–46, 51, 56–57, 70, 90, 92, 100, 104, 107, 109, 113, 120, 146–151, 153–154, 158
    Law of identity, 10–11, 52, 70, 79, 112–113
Ineraction, 10–11, 50, 52, 54, 56–57, 67, 102, 104, 108, 121, 123, 125, 127
Inside/outside, 51–52, 80–81, 90, 94, 97–100, 103, 108, 110, 145, 147, 150
Interpretation, 27, 69, 72, 74–76, 101, 153
Intuition, 1, 10, 20, 21, 23, 34, 131, 154

Knowledge
    Consumption, 55
    Self, 71

Laws of form, 89, 95

Memory, 71, 93, 103
Metarelation, 150
Metastructure, 61, 65, 67–70, 73, 77, 125
Mind
    Mind-reality, 4, 7, 17, 21, 69, 75–76, 156
    Collective mind, 41–42
Models, 35, 36, 127, 139, 140
Mysticism, 4, 83

Nothingness, 4, 83–86, 89–92, 94, 110, 131

Observer, 41, 62, 67, 80–81, 95–96, 99, 102–103, 107, 111, 120, 135, 144–152
Ontology, 17, 82–83, 120, 149
Origins, 7, 18, 23–27, 41, 49, 60, 70, 77, 79, 83–84, 89, 110, 128, 135, 152

Oscillation, 95, 98, 103–104, 106, 108, 135, 143, 146–147

Paradox, 95, 106, 135
Phenomenology, 18–19, 103, 112, 118, 133
Phenomenological reduction, 19, 59
Pre-objective, 1, 11, 13, 55, 70
Primordial, 1, 7, 10–14, 19, 25, 76, 79, 83, 92–93, 110, 112–113, 144, 148, 152–155

Recursive-hierarchy, 100, 102, 155
Reified universe, 1, 7, 16–18, 23–24, 44, 61, 77, 76, 79, 90, 98, 105, 109, 149, 152
Relational structure, 13, 79, 99, 109
Representation, 5, 7, 21, 29, 34–35, 62, 64–65, 123, 127, 139–140, 145, 148

Self organization, 135, 138
Self/non-self, 11, 103, 144, 146, 148
Semiotics, 4, 25, 35, 41, 46, 57, 76, 91, 97, 127, 153,
Symbol-grounding, 20, 23, 25, 28, 29, 52, 144
Set theory, 13, 102, 111
Solipsism, 18
Singularities, 10–11, 13–14, 62, 112, 120, 148–149
Socio-somatic-semiotic, 79, 82, 99, 123, 125, 142, 146
Spiral-like process, 18, 52, 79
Surface, 17, 34, 103–106, 108–109

Talmud, 27–29, 73–76, 85
Transgradience, 146

Utterance, 32, 51–53, 65, 67, 140–141

Vibration, 94–95, 104